Crossroads School Library
500 DeBaliviere Ave
St Louis MO 63112

Nuclear Proliferation

The Problems and Possibilities

Glenn Alan Cheney

Crossroads School Library
500 DeBaliviere Ave
St Louis MO 63112

An Impact Book

FRANKLIN WATTS
A Division of Grolier Publishing
New York London Hong Kong Sydney
Danbury, Connecticut

Frontpiece: The bomb being dropped on Hiroshima, August 6, 1945

Photographs ©: AP/Wide World Photos: 72 (Mark Avery), 107 (Ed Bailey), 68 (Richard Drew), 100 (Greg Gibson), 60 (Stringer), 27, 82; Corbis-Bettmann: 92 (Agence France Presse), 21 (Novosti), 11, 16; Department of Energy: 54; Gamma-Liaison: chapter openers, 4, 53; Los Alamos National Lab: cover; New York Public Library Picture Collection: 78; Reuters/Archive Photos: 80 (Aladin), 64 (Mikhail Chernichkin), 39 (Win McNamee); Sovfoto/Eastfoto: 51 (V. Yegorov/ V.Mastiukov), 28; Sygma: 96 (Kazakhstan Kamenogorsk); UN Photo: 34 (Y. Nagata), 25, 26, 31; UPI/Corbis-Bettmann: 117.

Visit Franklin Watts on the Internet at:
http://publishing.grolier.com

Library of Congress Cataloging-in-Publication Data
Cheney, Glenn Alan.
    Nuclear proliferation : the problems and possibilities / Glenn Alan Cheney.
    p. cm.—(Impact book)
    Includes bibliographical references and index.
    Summary: Discusses current policies governing the spread of nuclear weapons along with potential problems and possible outcomes.
ISBN 0-531-11431-7
1. Nuclear nonproliferation—Juvenile literature. 2.Nuclear weapons—Juvenile literature. 3.Radioactive substances—Juvenile literature. [1. Nuclear nonproliferation. 2. Nuclear weapons. 3. Disarmament. 4. Radioactive substances.]
I. Title.
JZ5675.c45 1999
327.1'747—dc21
98-4526
CIP
AC

©1999 by Glenn Alan Cheney
All rights reserved. Published simultaneously in Canada
Printed in the United States of America
1 2 3 4 5 6 7 8 9 10 R 08 07 06 05 04 03 02 01 00 99

Contents

Chapter 1	Worst Case Scenarios	6
Chapter 2	The Nature of the Beast	13
Chapter 3	The Initial Proliferation	24
Chapter 4	Treaties and Technology	38
Chapter 5	The Post-Soviet Nuclear Powers	57
Chapter 6	The Status of Proliferation	66
Chapter 7	The Rogue Regimes	79
Chapter 8	Deadly Traffic	90
Chapter 9	Nonproliferation Policies	105

Glossary	119
Source Notes	123
For Further Information	134
Index	138

The bomb being dropped on Hiroshima, August 6, 1945

Executive Order

Measures to Restrict the Participation by United States Persons in Weapons Proliferation Activities

By the authority vested in me as President by the Constitution . . . I, William J. Clinton, President of the United States of America, find that the proliferation of nuclear, biological, and chemical weapons, and of the means of delivering such weapons, constitutes an unusual and extraordinary threat to the national security, foreign policy, and economy of the United States, and hereby declare a national emergency to deal with that threat. . . .

William J. Clinton
The White House, September 30, 1993[1]

Chapter 1

Worst Case Scenarios

The worst scenario might start on a cold morning at a navy base on the coast of the Arctic sea. A young soldier sits on a crate outside a rusty sheet metal building, his Kalashnikov rifle across his lap. As a corporal in the Russian army, he is earning the equivalent of $31 a month. He had beet soup for breakfast, beet soup for lunch. When he hears a truck coming through the front gate without stopping, he hopes it's the one that brings bread. But it doesn't go to the mess hall. It comes straight to him and stops. Several men get out. While a few take a sledgehammer over to the door of the building, another offers the young soldier a deal he can't refuse. He can have his throat cut right then and there, or he can accept a $200 tip if he'll take a little walk for a while. The boy quickly understands how the truck was able to drive onto the base so easily. He accepts the money—not worthless Russian rubles but U.S. dollars—and strolls over toward the barracks. He can hear the men smashing the lock off the door, but what can he do? If they really want nuclear submarine fuel that bad, it's not his problem.

Or the worst scenario might start at the bazaar in the Nowy Targ Valley in Poland. Several gentlemen with foreign accents are browsing around a sizable pile of Stinger surface-to-air missiles that had disappeared during the war against Iraq in 1991. A box of hand grenades looks interesting, too, as does a case of AK-47 machine guns, relics of the war between the former Soviet Union and Afghanistan. But what these men are really looking for, they tell a certain merchant, is something radioactive. Some plutonium would be nice, or something in an enriched uranium. The merchant offers a small sample, a lead box containing a pellet from a fuel rod stolen from the Chernobyl nuclear power plant in Ukraine. The pellet is highly radioactive—not good enough for a nuclear bomb but quite adequate as a powerful poison. The shoppers look interested and ask how much more might be available and whether the seller would prefer to be paid Swiss francs or Burmese heroin.

Or perhaps the worst scenario might start in Asia when a border skirmish between two traditional enemies begins to look like a full-scale invasion. The ruler of the smaller country, knowing his country cannot resist the onslaught of a quarter of a million troops, orders his only atomic bomb loaded onto a specially equipped fighter-bomber. Although his country has never tested the product of their new nuclear technology, he announces to the world that, as long suspected, his country is indeed a nuclear power. If the invasion is not curtailed and reversed, he will launch an annihilation of biblical proportions. The minister of defense of the invading country announces that it, too, has nuclear weapons and will not hesitate to use them to reciprocate a nuclear attack.

Perhaps a similar scenario in the Middle East might threaten to involve the United States. A regional power invades an oil-rich neighbor. The invader is known to have been trying to develop an atomic weapon. But no one is sure whether it succeeded. The United States is called on to oust the invader and protect the oil fields of the region. Should American troops be deployed on a foreign battleground that could well turn into a nuclear target?

In a less likely but by no means impossible scenario, a helicopter hovers over the southern end of New York's borough of Manhattan. Inside, a woman holds a heavy, lead-lined box. Tossing it out of the helicopter, she watches it fall. Within two seconds, it explodes with the power of a hand grenade, releasing a burst of high-level radioactive waste that came from a nuclear power plant in Asia. Several pounds of cesium, uranium, plutonium, strontium, and other radioactive isotopes swirl outward in the downdraft below the helicopter. She flips open a cellular phone, dials a local radio station, and advises them to begin evacuating the Wall Street area. It's radioactive, she tells them, and it will probably remain radioactive for the next several years.

In another scenario, authorities in Los Angeles have detected abnormally high levels of radiation at many points throughout the city and its suburbs. It is especially heavy along freeways and in certain neighborhoods. Over the course of a few days, they track down several sources—gas stations that have been selling gasoline contaminated with radioactive materials. Further investigations trace it to an oil tanker that arrived from the Middle East with several million gallons of contaminated oil. The important question is not which country it came from but what to do about a city that is slightly, but perhaps dangerously, radioactive.

These scenarios are products of the imagination, but events quite like them have either taken place already or might well take place someday. Pakistan is believed to have prepared a nuclear attack against India.[1] North Korea is believed to have refined enough plutonium for at least one atomic bomb.[2] Until the Gulf War of 1990, Iraq was within a few years of building a bomb. A religious sect in Japan was looking for ways to use chemicals and radioactive materials as weapons and succeeded in an attack in the Tokyo subway.[3] Criminals in Russia, perhaps linked to international organized crime, have gained access to radioactive material, some of it pure enough to make nuclear bombs, some of it so toxic it could render a city uninhabitable.[4] And no one can rule out the possibility that atomic bombs have been stolen. Several smugglers have been caught with plutonium, uranium, cesium, and other dangerous isotopes.[5] Presumably, other smugglers have not been caught.

Nuclear proliferation is the spread of nuclear weapons and radioactive materials that could be used as weapons of contamination. The danger of nuclear proliferation has become so serious that U.S. President Clinton declared a national emergency, allowing the government to put special restrictions on certain products and exports.[6] U.S. Representative Tom Llantos, chairman of the House Subcommittee on International Security, International Organizations and Human Rights, said, "No danger to our national security is greater than the continuing spread of nuclear weapons to undeclared nuclear states. . . ."[7] Louis J. Freeh, director of the Federal Bureau of Investigation (FBI), said to the Russian Police College, "one criminal threat looms larger than the others: the theft or diversion of radioactive materials in Russia and Eastern Europe."[8] In a nationally televised speech, Senator Richard Lugar said that nuclear terrorism is "but one small step" away because of "grossly inadequate" control of radioactive materials in the former Soviet Union.[9]

These U.S. leaders have recognized a growing and very serious problem. The power of nuclear holocaust that was once the privilege of a handful of major nations is now within the reach of relatively undeveloped countries. Nuclear weapons may soon be available to terrorist groups that are not controlled by any government. The potential for destruction defies the imagination, and the solution is still beyond anybody's guess.

Since 1969, most countries have adhered to the Nuclear Nonproliferation Treaty (NPT), an agreement that countries without nuclear weapons will not try to acquire them. The terms of the treaty, however, make it easier for countries without nuclear weapons to buy the equipment needed to make them. Countries that have chosen to ignore the treaty have been making rapid progress toward developing a nuclear force.

Meanwhile, nuclear technology has advanced and become more readily available. The amount of fuel needed for a bomb has dropped to only a few kilograms. At the same time, the nuclear energy programs have increased the world's supply of radioactive resources from which weapons-grade nuclear fuel can be separated. As if the availability of recipe and ingredients wasn't enough, the situation has been aggravated by the breakup of the Soviet Union. As post-Soviet societies tumble toward bankruptcy and anarchy, their nuclear materials, equipment, technology, and personnel are finding their way into other countries.

The situation looks ominous for the twenty-first century. It will follow a century of terrible wars, exploding populations, breathtaking progress, religious strife, soaring technologies, devastated environments, booming economies, and a global communication system that has knit the world's countries into a virtual community. What will it all add up to if weapons of mass destruction are in

U.S. President Bill Clinton signing a comprehensive nuclear test ban treaty at the United Nations in September 1996. Clinton has made many efforts to restrain nuclear proliferation.

the hands of tyrants and terrorists? To a great extent, the next few years will determine how well the world will deal with the proliferation of nuclear weapons. The decisions and solutions of the immediate future will have repercussions for many years to come. Now, therefore, is the time to understand what is happening and what needs to be done.

The Nature of the Beast

Nuclear technology can lead to devastating consequences in two main ways. It can be used to build an atomic bomb, and it can create dangerous concentrations of highly toxic radioactive materials.

The Atomic Bomb

Building an atomic bomb is easier than it used to be. The key ingredients of the fuel, uranium-235 (U-235) and plutonium-239 (P-239), are more plentiful than ever. U-235 and P-239 are isotopes of the elements uranium and plutonium. Simply stated, isotopes are variations in the structure of the atoms of a given element. Radioactive isotopes are isotopes that are constantly giving off subatomic particles and electromagnetic waves. This activity of emitting particles and waves is called radioactivity.

Plutonium, which exists in nature in only trace amounts, is produced as the fuel of nuclear power plants decays. Through a difficult and expensive process, the plutonium can be separated from the spent nuclear fuel.

U-235, too, is rare in nature. Uranium in its raw form, when mined, is about 99.3 percent U-238. For use as a fuel in a nuclear power plant, that raw uranium is often enriched to 3 to 5 percent U-235, a concentration is known as low enriched uranium (LEU). Highly enriched uranium (HEU) is at least 20 percent U-235. HEU must be further enriched to over 90 percent U-235 to become weapons-grade fuel. P-239, normally enriched to 93.5 percent, is the most efficient bomb fuel as it yields the most energy for its weight.

The International Atomic Energy Agency (IAEA) reports that 1,000 tons (907 metric tons) of plutonium have been produced by the world's nuclear energy reactors.[1] About 22 pounds (10kg) of plutonium—and maybe less than half that amount—is enough to fuel an atomic bomb. A microscopic speck in the human lung is enough to cause cancer.

The United States and Russia produced about 1,750 tons (1,590 mt) of HEU and 230 tons (209 mt) of plutonium for weapons purposes.[2] The United States has also stored about 200 tons (182 mt) that are not needed for weapons.

Over the last fifty years, the United States has sent nearly 1 ton (0.9 mt) of plutonium to 39 countries, most of it for use as fuel or as a source of radiation. Germany received 1,280 pounds (518.1 kg), and Japan received 251 pounds (113.5 kg). About 2.2 pounds (1 kg) went to eleven countries, including Brazil, Colombia, Iran, Pakistan, Uruguay, and South Vietnam. Iraq, Czechoslovakia, the Philippines, Venezuela, and a few others received less than 2.2 pounds (1 kg).[3]

Neither plutonium nor highly enriched uranium is easily produced. Their production involves a series of high-technology processes which are really industries unto themselves. One method is to take raw uranium and refine it directly to HEU, a big step that calls for a rather sophisticated, high-tech factory processing a large amount of uranium. It takes months or years to purify enough fuel for a bomb.

The more common method of producing fuel for a bomb is to separate U-235 or P-239 from spent nuclear fuel.

As the fuel in a nuclear power plant gives off energy, it decays into radioactive isotopes of many different elements. Among them are U-235 and P-239, which are often separated to use as bomb fuel or as fuel for other nuclear reactors. Most of the spent fuel that is not separated for these uses will probably end up as nuclear waste. Obviously, this enrichment process requires a large supply of spent nuclear fuel and a sophisticated facility for separating the desired isotopes.

Nuclear fuel enriched to weapons-grade will be the core of a bomb. To explode, the core will have to reach critical mass. Critical mass is a concentration of radioactive material condensed to a certain point. At that point—the point of critical mass—neutrons shooting out of the radioactive atoms collide with and split other atoms. When those atoms break up, they release more neutrons, which split more atoms. This chain-reaction process is called fission. Nuclear fuel, in a bomb or the reactor of a power plant, is fissile material.

Just piling the fuel in a heap will not produce much of a fissile explosion, though critical mass could be reached. As soon as the chain reaction begins, however, the released energy scatters the fuel. The critical mass is then broken up, and the chain reaction stops.

So, to make an atomic bomb, it is necessary to condense a quantity of fuel to critical mass suddenly, in the briefest fraction of a second. The more precisely and quickly the fuel is condensed, the bigger the chain reaction, the bigger the release of energy, and the bigger the explosion. Inevitably, much of the fuel will not have time to react before the explosion hurls it away.

The trick to making an efficient bomb, therefore, and to determining the amount of fuel needed to make it explode, is the firing device that compresses the fuel into critical mass. Either of two techniques will work.

The implosion technique, which is used in most warheads deployed by the United States, uses high explosives to compress the fuel. The primary explosive material forms a sphere around the core. Electrical charges ignite several points of a primary explosive at precisely the same moment. The explosion must push uniformly and simultaneously inward on the core, condensing it to critical mass. The bomb that destroyed Nagasaki was an implosion bomb using plutonium fuel.

The gun-assembly technique slams together two masses of HEU. Once combined, they reach critical mass. In one version, a small detonator shoots a bullet of fuel into the core of the bomb. The bomb detonated at Hiroshima was of the gun-assemby type.

The "Little Boy" type of bomb, which was detonated over Hiroshima in World War II

Another type of atomic bomb, the fusion bomb, also called a hydrogen or thermonuclear bomb, uses fissile material to produce a fusion explosion. In a certain way, however, a fusion bomb is the opposite of a fission bomb. Rather than splitting atoms, a fusion bomb uses atomic energy to fuse atoms together. A primary explosion creates a fission explosion, which then creates a fusion explosion. The resulting release of energy can exceed that of a fission bomb.

Step one, then, in building a bomb is to acquire unenriched nuclear material—spent nuclear reactor fuel or a large supply of natural uranium. Step two is to master the technology needed to enrich it. Of course the direct acquisition of enriched, weapons-grade fuel would make these first steps unnecessary. Step three is to build a firing device unless, of course, such a device can be acquired somewhere.

None of these steps is quick, easy, or cheap. As explained in a later chapter, the specialized equipment and radioactive materials are monitored or controlled by international treaties. Taking these steps in secrecy makes the process even slower, more expensive, and more complicated.

Two steps remain. The owner of the bomb needs a control system to make sure the weapon isn't used until so desired. The owner also needs a delivery system.

The control system can be as sophisticated, complex and expensive as the war rooms of the United States and Russia, or as simple as a padlock on a door. Since the possessor of an atomic bomb holds tremendous power, at least as a threat, control systems are normally very tight. Ideally, only a few people have the code or the key or the other device or bit of knowledge that will allow a weapon to be delivered or detonated. Assuming that a given weapon exists only to deter an attack, the looser the control over a weapon, the more dangerous it is. One of the main dangers of nuclear proliferation is that unstable governments may acquire a bomb yet be unable to control the military people who hold it.

The delivery system can be a bomber, a missile, or, in the case of smaller "tactical" warheads, artillery. An atomic bomb can also simply be planted as a mine. Theoretically, a bomb can be smuggled across a border as cargo in an airliner or on a freight ship. A small speedboat could carry one to shore from a freighter or submarine. It could be carried in a heavy but portable backpack.

Atomic Explosions

Nuclear explosions produce damage in three ways. The explosion itself produces a tremendous shock. The sudden expansion of air sends a shock wave out from the explosion at thousands of miles per hour. At a distance of thousands of yards, the shock wave can knock down steel buildings and hurl heavy vehicles through the air. At farther distances, debris, especially glass, shoots through the air like bullets.

An atomic explosion also releases an astounding amount of heat. For a brief instant, the center of the explosion is many times hotter than the center of the sun. The quick flash of light is powerful enough to instantly blind anyone looking in that direction. The heat rushes outward in a huge fireball that rises into the sky. Vaporized earth and other materials shoot into the sky, then condense and fall to form the characteristic mushroom cloud. The fireball ignites objects miles away. The fires are so widespread that they become firestorms, sucking up vast amounts of air, producing an in-rushing wind that brings in yet more flammable material. The firestorm phenomenon also occurred during bombing raids with non-atomic explosives during World War II (1939–1945). In fact, the fire-bombings of Tokyo, Dresden, and other cities killed more people immediately than the atomic bombings of Hiroshima and Nagasaki. Radioactive fallout, however, caused deaths long after the atomic explosions. The radioactive materials of the warhead and those produced by the chain reaction of the explosion may have killed almost as many people as the explosions themselves.[4]

Radioactive fallout tends to consist of isotopes of cesium, strontium, iodine, plutonium, uranium, and a host of other elements. Once the heat of the explosion has raised them into the upper atmosphere, dangerous concentrations can be carried hundreds of miles.

People exposed to high doses of radiation are very likely to suffer radiation poisoning, which can result in death within days or weeks. People receiving lower doses stand a chance of developing cancer or suffering damage to the DNA of their reproductive organs. These lower doses can occur slowly from radioactive particles, called radionuclides, that have been swallowed or inhaled. If radionuclides stay in the body, they continue to damage bones, blood, organs, and the immune system. Cancer, leukemia and immunological problems are common results.[5] These effects may take years to occur. Scientists are still researching the effects of long-term, low-dose exposure to radiation.

Radioactive Contamination

Atomic explosions are not the only source of highly toxic radioactive material in the environment. These materials are widely present in industrial nations, and they are contributing to the general quality of life. Nuclear reactors at electric power plants use and produce thousands of tons of radioactive materials. Radioactive materials are also found in medical and scientific instruments, measuring equipment and even smoke detectors and exit signs.

The benefits of radioactive materials have so far been of tremendous help to society. As long as the radionuclides are contained within protective shielding, such as lead and concrete, they do no harm. Some radionuclides need even less protection. If, however, radioactive isotopes end up in the environment, near or inside the human body, they can cause severe health problems, some of which can be deadly.

The danger created by contamination of the environment depends on various factors. Some radioactive isotopes are more dangerous than others. Plutonium-239 (P-239), cesium-137 (Cs-137), strontium-90 (Sr-90), and iodine-131 (I-131) are among the most dangerous. Some isotopes, such as I-131, decay and disappear relatively quickly and are no longer dangerous; others, such as P-239 and U-235, continue to be dangerous for hundreds, thousands, or millions of years.

Contamination can occur by accident, such as after the 1986 explosion and fire at the Chernobyl nuclear power plant in Ukraine. An estimated 8,000 people may have died within ten years of the accident, and tens of thousands may be suffering health problems.[6] In an abandoned medical building in Goiânia, Brazil, demolition workers found a capsule in a radiotherapy machine. The capsule contained a salt of cesium-137. Over the next few days, the container changed hands a few times. Four people died of radiation poisoning, and another 249 were contaminated and are at risk for grave illness. Several houses had to be demolished and buried, and a great deal of soil had to be taken away for disposal as toxic waste.

A helicopter taking radiation measurements over Chernobyl after an accident at that plant in 1986

Contamination can happen on purpose, too. It is not hard to imagine ways of using radioactive materials as a weapon. A substantial amount in a city's drinking water could contaminate everything and everybody that the water touches. Contamination of a region's heating oil or gasoline could spread radioactivity over a vast area through furnace chimneys and car exhaust pipes. A fluid laced with radioactive material sprayed over an area by plane could contaminate the place for decades. A small, non-nuclear bomb packed with a few pounds of spent nuclear fuel could contaminate a few city blocks. The explosion itself would be no worse, but the effects of the contamination could be devastating.

In these scenarios, even slight contamination could cause catastrophe. A local population could panic. Real estate values could plummet. Industries could shut down. Transportation routes could be permanently blocked. The shock to stock markets could cause financial ruin. A single act of atomic terrorism could truly terrorize the world.

Controlling Radioactive Materials

In the United States and most other countries, radioactive materials are carefully, though by no means perfectly, controlled. Controls are stronger for the more dangerous materials. Because they are highly radioactive and can be used as bomb fuel, plutonium and HEU are the materials most carefully controlled and accounted for. Still, in the United States, a few tons of these materials have inexplicably disappeared. In the republics of the former Soviet Union, no one is certain how successfully these and other materials have been protected. This book explains these and related problems in later chapters.

As new technology uses radioactive materials for good purposes, the quantity of such materials in the world increases. At the same time, as more countries build nuclear power plants, more radioactive fuel must be produced, more radioactive waste must be disposed of, and more radioactive material must be stored and transported. All these activities involve some risk of loss or accident. In some countries, spent fuel from nuclear power plants is directly dedicated to production of bomb fuel, and some countries may be building nuclear power plants, not because they need the electricity but because they need the fuel. Again, this is explained in more detail later in this book.

The incident in Goiânia, Brazil, demonstrated how easily and innocently radioactive materials can fall into the wrong hands. As nuclear materials become more and more a part of useful technologies, the risk of the proliferation of radioactive fuels and materials increases proportionately. An underground market for such materials has developed, and the buyers are most certainly not in the market with good intentions.[7]

Nuclear proliferation, then, is happening in different ways. The production of radioactive materials has increased over the years, and more countries are involved in that production. At the same time, more countries are acquiring the technology needed to build atomic weapons. The fuel for those weapons is becoming more plentiful and possibly more accessible. The world has never had more control over nuclear weapons and materials, yet never have nuclear products been more widely, and illegally, available.

Chapter 3

The Initial Proliferation

World War II ended when the United States detonated atomic bombs over the Japanese cities of Hiroshima and Nagasaki. The Hiroshima bomb yielded about 13 kilotons, that is, it had the explosive power of 13,000 tons (12,000 mt) of TNT. It killed 45,000 people immediately. (Some estimates put the toll much higher.) Another 19,000 died within the next four months. Deaths due to cancer and leukemia have continued since then, bringing the total to an estimated 136,000. The 23-kiloton bomb that destroyed Nagasaki killed 22,000 on the first day. Another 17,000 died soon thereafter, and a total of 64,000 have died in all.[1]

The sudden appearance of nuclear weapons in the world, and the demonstration of their power, ended an era of relative simplicity and opened the way to war of unimaginable horror. Shortly after the end of World War II, the United States proposed a plan to prevent the proliferation of these weapons to other countries. The Baruch Plan, named after Bernard Baruch, a U.S. representative to the United Nations (UN), proposed gradually reducing the U.S. monopoly on nuclear weapons and sharing nuclear technology, for peaceful purposes, with the whole world. The newly formed United Nations would oversee all production of nuclear materials, beginning with the mining of uranium. Addressing the UN, Baruch warned that "terror is not enough to inhibit the use of the atomic bomb. The terror created by weapons has never stopped man from

These survivors of Hiroshima suffered serious injuries and faced a lifetime of illness.

Smoldering ruins the day after the bomb was dropped on Hiroshima

employing them." He went on to say that "for each new weapon a defense has been produced in time. But now we face a condition in which adequate defense does not exist."

Baruch's plan, however, was never put into action. The Soviets objected to the verification system, and the U.S. refused to destroy its weapons before an international authority was established.

The Cold War

The Soviet premier at that time, the tyrannical Joseph Stalin, refused to let American technology and military power dominate the world.[2] World War II had cost his country 20 million lives. Justifiably, he felt compelled to build an army powerful enough to repel any invasion. The Soviet Union launched an all-out effort to build an atomic arsenal. Predictions of how long it would take the Soviets to build their first bomb ranged from five to twenty years. Americans hoped something would stop them before then.

In early 1946, Stalin renounced the World War II alliance between his country and the United States. He also stated that war between his communist country and the capitalist countries was inevitable. Later that year, British Prime Minister Winston Churchill said that the Soviet Union had lowered an "Iron Curtain" between itself and the rest of the world. In effect, the Soviet Bloc, which included several satellite countries along the Soviet border, and the Free World of Western Europe and North America, had, in effect, declared themselves enemies. Thus began the Cold War.

Bernard Baruch, U.S. representative to the United Nations, tried to control nuclear proliferation after World War II, but his plan was never accepted.

Following World War II, Soviet Premier Joseph Stalin built a huge nuclear arsenal for his country, hoping to prevent invasion from other countries.

In April 1949, the North Atlantic Treaty Organization (NATO), a military and economic alliance of several non-communist countries, was established to prevent Soviet military expansion. Then, on August 29, 1949, to the surprise of the whole world, the Soviet Union tested its first atomic bomb.

Within a year, the Korean War broke out. The United States and other Western countries helped defend South Korea. The communist People's Republic of China helped defend North Korea. The line that divided North and South Korea, the 38th Parallel, became the battlefront between the communist powers and the Free World. The fighting was fierce. When the U.S. forces had trouble holding back waves of Chinese soldiers, the American military considered the use of atomic weapons. President Dwight Eisenhower hinted, through diplomatic channels, that the United States was considering using nuclear weapons to repel the Chinese invasion. China took note, of course, and dedicated itself to the development of its own bomb.

In October 1952, Britain, with the help of the United States, tested its first atomic bomb. On November 1 of that year, the United States exploded the first hydrogen bomb. Three days later, Dwight D. Eisenhower was elected president. On July 27, 1953, the Korean War ended. In 1955, the Soviet Union detonated its first hydrogen bomb.

The Cold War was being waged with the loudest saber-rattling the world had ever known. As if vying to show the world who could build the biggest bomb, the two superpowers flaunted their megatonnage. Average-sized bombs had a megaton of power—the equivalent of 1 million tons (900,000 mt) of TNT. The most powerful hydrogen bombs yielded over more than 15 megatons. The news media commonly referred to the power of the bombs in multiples of the 13-kiloton Hiroshima bomb. The more powerful bombs were hundreds of times more powerful than the one that had killed over 125,000 people in Japan.

During the 1950s, the United States produced nuclear warheads at a remarkable rate. By the end of the decade, almost 20,000 warheads were produced in a three-year period—a rate of 75 per day, building the national arsenal to 22,000 warheads in 1961. By 1965 the total stockpile had reached 32,400, a total yield of 15,152 megatons.[3]

The Soviet arsenal developed more slowly but reached higher totals. In 1987, the Soviet Union had an estimated 45,000 warheads, though apparently many of them were old, antiquated weapons.

France joined the "nuclear club" in 1960 with a device detonated in the Algerian Sahara. Critics believed that France developed its bomb only so that it would never be considered a second-rate country. It was a bomb of status more than defense.[4] Still, the advantage of a nuclear arsenal, however small, is that just a few bombs can inflict more damage than an enemy is willing to accept. Though France might never win a nuclear war against the Soviet Union, it developed enough of an arsenal to prevent one. Also, France was not sure that the United States would come to its rescue if it were engaged in a nuclear war.

Never forgetting that its forces had almost been the target of a U.S. nuclear attack in Korea, China finally exploded its first bomb in 1964. Just as Britain had used American technology to develop a bomb, China used Soviet technology. The Soviets soon regretted their nuclear aid. Soviet-Chinese relations soon deteriorated. When China tested its first bomb, the two countries became the first nuclear powers to share a border.

Nikita Khrushchev increased nuclear weapons production in the Soviet Union during the 1960s.

For the next ten years, the United States, the Soviet Union, Britain, France, and China would remain the only declared members of the nuclear club. During that time, the two superpowers raced to maintain superiority in bombs, missiles, bombers, and submarines. The word "overkill" was introduced to the English language. Overkill is the destructive nuclear capacity exceeding the amount needed to destroy an enemy. By cynical definition, it is the ability to kill a given population more than once. The leader in the arms race was the country that could kill the other country's people the most times. Between the two of them, the Soviet Union and the United States eventually had the equivalent of 6,000 pounds (2,700 kg) of TNT for every human being in the world.

In 1962, U.S. spy planes discovered the Soviet Union was installing missiles, possibly with nuclear warheads, in its communist ally, Cuba. In a showdown of strength, the United States prevented further delivery of missiles and negotiated the withdrawal of all such weaponry from Cuba. Apparently Soviet Premier Nikita Khrushchev blamed his loss of strategic position—not to mention loss of face—on inferior atomic firepower. He increased his country's weapons production, and the United States responded in kind.

With these tremendous intercontinental atomic arsenals aimed at each other, the two countries were locked in a standoff known as MAD—mutually assured destruction. If either side started a nuclear war, the other side would retaliate in kind, and massively. Both sides, and perhaps the whole world, would surely be destroyed. As mad as the strategy sounded, however, it did provide a good reason not to launch the first missile. In fact, it may have helped the two superpowers refrain from directly facing each other on a battlefield. A war with conventional weapons might easily have "gone nuclear." The battles between the superpowers, it turned out, would be fought indirectly in countries such as Vietnam, Nicaragua, Afghanistan, and Angola. In these countries, forces backed by the Soviet Union fought against forces backed by the United States.

In the mid-1980s, the MAD balance of power reached its peak at a total of about 70,000 nuclear warheads between the two superpowers.[5] The warheads were deployed on intercontinental missiles, long-range bombers, and nuclear submarines. Some of the missiles carried several Multiple Independently-targeted Reentry Vehicles (MIRVs), which could separate from the missile and hit different targets. As an example of the power in these weapons, one U.S.

Poseidon submarine could destroy more cities than the Allied forces destroyed in all of World War II. Its 16 missiles carried more explosive power than a million World War II B-17 bombers. And yet the explosive power of that one submarine represented less than 1 percent of the megatonnage in the U.S. arsenal. The Poseidon submarines have since been replaced by the even more powerful Trident and Seawolf submarines.

Mutually assured destruction may have tended to prevent the use of nuclear weapons. At the same time, however, the overkill capacity of the opposing nuclear force made retaliation difficult, if not impossible. A massive attack with tens of thousands of warheads could destroy the enemy once and for all. Since missiles could fly north from either country, cross the polar region, and reach the other nation's territory in under an hour, even quick notification of an attack would give the other country too little time to decide to counterattack. If a political or military situation became too tense, therefore, both countries would be tempted to go ahead and launch rather than wait to see if the other side launched. Their nuclear forces were on a hair trigger. Both sides had a "launch-on-warning" strategy. If they saw an attack coming, they would retaliate before the first bombs exploded.

Needless to say, that disaster never came to pass. The two superpowers were able to maintain control over their nuclear forces. Their governments were stable enough for each side to know who was in power on the other side and what their international policies were. They found non-nuclear ways to confront each other and to share power.

The arms race, the Cold War, and worldwide fears of nuclear holocaust reached a peak in the early 1980s. Ronald Reagan was president of the United States, and

Mikhail Gorbachev became president of the Soviet Union in 1985 and made strides toward economic and political reforms as well as arms reductions.

Leonid Brezhnev was premier of the Soviet Union. Despite some treaties that limited the numbers of certain types of warheads and delivery systems (see chapter 4), tensions were high. President Reagan began a massive military buildup in hopes of driving the Soviet Union bankrupt as it tried to keep up. He began plans for building a missile defense system that would be based in outer space. The Soviets, balking at a technology they could never match, threatened to destroy any such system before it could be completed.

The tension and seeming insanity began to ease in 1985 when Mikhail Gorbachev became president of the Soviet Union. Gorbachev recognized that his country would never prosper under its regime of suppressed freedom, controlled economy, and ongoing conflict with the Western countries. His solutions were policies called perestroika (economic and political reforms) and glasnost (more freedom of expression by individuals and the press). He also made sweeping proposals for arms reduction.

The Soviet Union had not known such freedom since its birth in 1922 or, for that matter, in the pre-Soviet centuries of rule under the czars. The country quickly broke down into something close to economic and political chaos. As it tried to shift from the centrally controlled economy of communism into a system more like free enterprise, the economy faced a situation worse than the Depression that affected the Western world in the 1930s.

For many years, the Soviet Union had controlled several formerly independent republics, such as Latvia, Lithuania, Estonia, Ukraine, Belarus, Armenia, and Kazakhstan. None of them liked Soviet control. As political domination loosened and the economy provided for fewer of anyone's needs, these republics began declaring their independence. During 1991, the Soviet Union came untied, and by the end of the year, by vote of the Soviet parliament, the Soviet Union ceased to exist. Mikhail Gorbachev resigned on December 25 of that year.

Of the fifteen new countries that came out of the Soviet Union, four were nuclear powers. Russia held the most weapons by far. Ukraine, with 1,256 nuclear weapons systems, became the world's third-largest nuclear power. Kazakhstan had 1,410, and Belarus had 72.

Among them, only Russia has remained a nuclear power. Belarus quickly gave up its weapons, but Ukraine and Kazakhstan held out. After long negotiations, they agreed to sign the Nuclear Nonproliferation Treaty and to surrender their weapons in exchange for promises of foreign aid and the security of their borders. The process of destroying the weapons or transferring them to Russia, however, would take several years. As of mid-1998, Kazakhstan still had not signed the treaty.

The new governments were very weak, disorganized, inexperienced at democracy, and quite corrupt. Boris Yeltsin succeeded Gorbachev as president of Russia. His struggles with the Russian parliament demonstrated how Russian government was barely able to govern itself, let alone the vast reaches of the world's largest country. The other post-Soviet republics had similar problems adjusting to the liberties of democracy.

A New Nuclear Threat

With the collapse of the Soviet Union, the Cold War was over. Soviet troops—now Russian troops—began pulling out of East Germany, which had been dominated by the Soviet Union. The United States took its bomber fleet and intercontinental missiles off alert. The two superpowers—if Russia might still be included as one—agreed that they were no longer enemies. For a short time, it looked as if the world might soon become free of the nuclear threat that had hung over civilization for almost half a century.

Almost at the same time, however, new nuclear threats emerged. Smugglers in Europe were caught with small amounts of very deadly radioactive materials. In 1991, Iraq was found to have been close to developing an atomic weapon. In 1993, North Korea was suspected of purifying enough plutonium for one or two bombs. In 1994, a journalist published an article revealing that Pakistan had almost sent a jet fighter to drop an atomic bomb on New Delhi, India, though Pakistan was still claiming not to have a bomb.[6] Iran was suspected of working on an "Islamic Bomb," and Libya was in the market for nuclear fuel. By late 1994, nuclear technology had reached a point where as little as 2.2 pounds (1 kg) of plutonium might suffice to fuel a high-tech one-kiloton bomb.[7] In 1998, India and Pakistan surprised the world with two series of underground nuclear tests.

The nuclear monster had not been caged. It was still loose, and in fact it seemed that its deadly seeds had been scattered around the world and were sprouting one by one.

Chapter 4

Treaties and Technology

The balance of mutually assured destruction between the United States and the Soviet Union was dangerous enough. Dozens of countries engaging in a complex balance of terror would be incalculably worse. Regional skirmishes that used to involve tank and troop movements could quickly turn into nuclear war. The presence of nuclear weapons in too many hands greatly reduced the ability of the superpowers to "manage" crises within their spheres of influence.

The world feared that if the production of nuclear weapons became widespread, tyrannical governments could all the more easily acquire bombs of their own.

Weak or unstable governments might import such weapons but then lose control of them to their own military leaders or to a rebel group. Atomic weapons would be especially dangerous in the hands of governments led by dictators or fanatics. Unfortunately, these are the governments that tend to be most interested in acquiring nuclear weapons.[1]

Terrorists might also be especially interested in getting their hands on a nuclear bomb. This would be all the easier if nations that support terrorism had a bomb to contribute to the cause. Moreover, it would be hard to stop terrorists through threat of reprisal, given the suicidal nature of many bombings.

Russian President Boris Yeltsin, U.S. President Bill Clinton, Ukraine President Leonid Kuchma, and British Prime Minister John Major (seated left to right) signing the Nuclear Nonproliferation Treaty during a summit in 1995.

Nuclear Nonproliferation Treaty

It is far easier to prevent nuclear anarchy than to correct it after it happens. In 1968, therefore, the United Nations devised the Nuclear Nonproliferation Treaty (NPT). The treaty went into effect in 1970 for a twenty-five-year period. Eventually, most of the world's countries would sign.

The agreement was lopsided but better than nothing. Acknowledging reality, it let the five current nuclear powers (the United States, the Soviet Union, France, Britain, and China) keep their nuclear weapons as long as they tried to negotiate their elimination. All other member-states agreed not to acquire such weapons or even work on developing weapons technology.

The United Nations' International Atomic Energy Agency (IAEA) was appointed to serve as the watchdog of the agreement. All member-states agreed to allow the IAEA to monitor the production and movement of nuclear materials that could be used to build a weapon. (The five nuclear powers are exempted from observations of activities relating to nuclear weapons.) All facilities containing nuclear materials must be declared as well. (Again, the five nuclear powers are exempted.) In this way, if any country begins to build an atomic bomb, other countries will know and can take appropriate action. This detection is meant to work as a deterrent. The inspections are supposed to make countries less suspicious of their enemies, and in most cases, they succeed.

The NPT put only one significant restriction on the five members of the nuclear club. In exchange for being allowed to retain their nuclear weapons, they agreed to "pursue negotiations in good faith on effective measures relating to cessation of the nuclear arms race." They are supposed to work toward the goal of "general and complete disarmament."

A handful of key nations did not sign the NPT. South Africa signed only in 1994, at which time it declared that it had developed an atomic arsenal but had already disposed of it. Israel, always fearing attack from its Arab neighbors, did not sign and is believed to have 200 or more such weapons. India never signed the NPT, yet in 1974 it conducted its first nuclear test, and in 1998 it conducted five more. The Institute for Science and International Security estimates that India has stockpiled enough fuel for 74 weapons.

Pakistan, on India's northwest border, also declined to sign and, a few weeks after India's 1998 tests, surprised the world with several underground nuclear detonations.

In May 1995, the 149 signers of the NPT (and most other countries) met at the United Nations and extended the treaty indefinitely, making it permanent. As the countries came together, however, they were not in total agreement. Some nations said that if non-signers such as Pakistan, India, and Israel had weapons and no inspection, then other nations had the right to the same weapons.

Egypt specifically complained about Israel's alleged arsenal. Without admitting to an arsenal, Israel said it could not sign the NPT if two long-standing enemies, Iraq and Iran, had nuclear programs. Iraq and Iran, of course, denied having any such programs.

Other critics pointed out that the IAEA was not capable of detecting secret nuclear programs such as those in North Korea and Iraq, which had escaped detection for so many years.

Still others complained that the countries with nuclear technology were not sharing it equally and completely, as they were required to do under the treaty. Japan, for example, was given access to plutonium and related technology, while North Korea was not.

The most widespread argument against making the treaty permanent was that the superpowers had not made sufficient progress in disposing of their weapons. Despite a twenty-five-year-old promise to eliminate their weapons, they hadn't even come close. The U.S. arsenal was equal to its 1959 level, and the Soviet level was close to its 1977 level. Under Strategic Arms Reductions Treaty (START) agreements reached in 1991 and 1993, Russia and the United States had promised to reduce their strategic arsenals to about 3,500 warheads each. The U.S. Senate did not ratify the START II treaty until 1996, and as of the middle of

1998, Russia had not yet signed. Nonetheless, President Clinton and Russian President Yeltsin met in 1997 and agreed to begin START III negotiations as soon as the Russian parliament had agreed to START II. START III would try to lower each country's arsenal to 2,000 to 2,500 warheads by the year 2007. Many nations, however, felt the START reductions were too slow and did not go far enough.

After much debate, the NPT was made permanent when more than 170 countries signed an agreement on May 11, 1995. Israel, Pakistan, and India continued to refuse to sign. Also not signing were Brazil, Chile, Cuba, Oman, and the United Arab Emirates.

The agreement continued to call for "the determined pursuit by nuclear-weapon states of systematic and progressive efforts to reduce nuclear weapons globally, with the ultimate goal of eliminating those weapons." The nuclear-weapon states also promised to stop all testing of nuclear weapons in 1995. All signers agreed to annual reviews of their compliance with the terms of the treaty.[2]

The extension treaty specified twenty "Principles and objectives for Nuclear Non-Proliferation and Disarmament," as follows:

- Countries that have not signed the NPT should do so as soon as possible.
- Every effort should be made to make all parts of the treaty work.
- Nuclear-weapon countries should negotiate toward complete nuclear disarmament.
- Nations should negotiate a "Comprehensive Test Ban Treaty" to end all testing of nuclear weapons. Nations should also negotiate a treaty to end the production of fissile materials that can be used for weapons.

- "Nuclear-weapon-free zones," in which all nuclear weapons are prohibited, are a good idea that would enhance global and regional security.
- The development of nuclear-weapon-free zones should be a high priority, especially in regions of tension, such as the Middle East.
- All nations should cooperate in respecting nuclear-weapon-free zones.
- Legally binding agreements should be made to assure non-nuclear-weapon countries that nuclear weapons will not be used against them.
- Nothing should be done to undermine the authority of he IAEA and its safeguard system.
- Countries that have signed the NPT but not signed safeguard agreements should do so without delay.
- IAEA safeguards should be regularly assessed and evaluated. The IAEA's ability to detect undeclared nuclear facilities should be improved. Nations that have not signed the NPT should agree to IAEA safeguards anyway.
- Before selling fissile material to a non-NPT nation, the seller should require acceptance of all IAEA safeguards and inspections. The buyer should also sign an agreement not to develop nuclear weapons.
- Fissile material transferred from military to nonmilitary use should be put under IAEA safeguards.
- Research on, and the use of, peaceful nuclear energy is an "inalienable right" of all NPT countries.
- The "fullest possible exchange of equipment, materials, and scientific and technological information for the peaceful uses of nuclear energy should be fully implemented."

- Activities to promote the peaceful use of nuclear energy should give priorities to non-weapon NPT states. The needs of developing countries should be taken into account.
- NPT countries should let their nuclear-related export control be known to all other countries.
- All countries should maintain the highest possible levels of nuclear safety, including waste management, accounting for nuclear materials, and the protection of nuclear materials.
- Attacks against peaceful nuclear facilities jeopardize safety and should require application of international law and UN action.

Why the NPT Works ... Almost

The NPT has several factors working in its favor. The primary reason it has been successful is that none of the nuclear powers see any advantage in sharing their power with other nations. Instead, the superpowers have offered their "nuclear umbrellas" to some of their allies, theoretically threatening nuclear reprisal for nuclear attack. This factor, however, has little to do with the NPT. Regardless of the treaty, most of the countries with nuclear weapons have no strategic reason to help other countries develop them.

Another significant factor supporting the NPT was that by 1968 the world already had a tradition of non-use of nuclear weapons. The first two attacks at the close of World War II were, to date, the last two attacks. Nuclear powers have confronted each other indirectly in regional wars in Vietnam, Korea, Central America, the Middle East, and Afghanistan, sending troops and arms to support friendly governments. Despite grueling wars, neither side launched its ultimate weapon. By the 1990s, when it became technically possible for several less developed countries to acquire atomic bombs, half a century of non-use had made nuclear weapons seem less necessary. The actual use of a nuclear weapon on a battlefield has become increasingly taboo. The longer the world goes without conducting war with atomic bombs, the less necessary those bombs seem.

Of course the term "use" here means actual detonation in war. The truth is, atomic arsenals have often been used for political and diplomatic purposes. Even when they aren't mentioned in negotiations, everyone knows who has them and who doesn't. Like poker players with big piles of chips, countries with nuclear weapons have an advantage in negotiations.

Factors Against the NPT

Though the NPT has been generally successful, several factors work against it. Regional conflicts have continued despite the end of the Cold War. The nature of the conflicts is different, however. Now, without a superpower to back it up, a small country might see a need for some other "equalizer." A nuclear weapon might make a small army seem very big. And of course when one country in a region has such a weapon, its neighbors and traditional enemies will naturally feel the need to build a balance of power. The 1998 tests by India and Pakistan are perfect examples of this.

Meanwhile, the proliferation of warhead delivery systems, such as missiles, planes, and artillery, has made the acquisition of a bomb more attractive. In the past, though a country might have a bomb, it may not have had a way to explode it far from home. The balance of power, then, would have been more strictly regional. Modern missile technology and the sale of second-hand fighters and bombers to less developed countries make it easier for them to deliver a warhead outside their region. A single new nuclear power, therefore, will compel all the other countries to acquire equal power.

Delivery has also been simplified by technology that has made it possible to build smaller atomic weapons. They fit on smaller missiles and planes and can be more easily smuggled into a target country by boat, in a light plane, or even in a box on a commercial airliner. That same technology also makes it easier for atomic weapons to be transported among illegal buyers and sellers.

Other technologies have made it easier to develop atomic weapons. With modern "dual-use" technologies such as computers, metal-forming equipment, and medical diagnostic machines, countries can assemble secret facilities. No one would suspect why they bought the equipment.

Technology has also made it simpler to hide the fact that a country has built a bomb. A country no longer needs to test a bomb to see if its design works. Sophisticated computer models tell bomb designers all they need to know.

Developing a nuclear weapon has also become much less expensive. South Africa was able to develop the entire industry necessary to produce the fuel and components for less than $200 million. Fewer than 400 people were involved at any one time.[3]

As discussed later in this book, the end of the Cold War may be making proliferation easier. The collapse of the Soviet Union left Russia with too little control over its weapons and radioactive materials. Weak governments and the poorly guarded borders of Ukraine, Belarus, Kazakhstan, and Russia itself have made it easier for nuclear materials to find their way outside of the old Soviet borders. Likewise, unemployed nuclear scientists and technicians have been seeking employment in other countries. When they go, they take nuclear technology with them.

In one way, the NPT makes it easier for nations to acquire the fuel and equipment they need to build a bomb. The treaty specifically promotes the sharing of peaceful nuclear technology. That technology makes it easier, for example, to recover plutonium from spent power plant fuel. It makes it easier to buy and use the equipment needed for nuclear work.

Inspection is also a problem. The NPT can be no more effective than IAEA inspectors. Without major funding, these inspectors cannot possibly watch over all nuclear facilities, stockpiles, movements of nuclear materials, and ongoing processes involving plutonium. The problem worsens as plutonium becomes a more common commercial product.

IAEA inspections are especially ineffective in the cases where they are most needed—at facilities dedicated to bomb development. The agency has neither the technical nor legal ability to seek out and inspect secret facilities or undeclared stockpiles. For this reason, Iraq and North Korea were able to make a lot of progress toward building bombs before the IAEA suspected them. As explained in chapter 7, they found ways to secretly buy special equipment and to operate secret facilities.

The NPT also indirectly rewards countries that do not sign the treaty. Their nuclear facilities are safe from inspection, while those who sign must allow inspections.

The fundamental weakness of the NPT is that there are no "nuclear police." No international force has the authority to invade a country, seize its illegal nuclear weapons and haul its nuclear felons off to jail. If a particular government decides to ignore the principles of the NPT and to develop an atomic bomb, the international alternatives are few. The world is going to have to develop new treaties and technologies to deal with this new kind of crime.

Other Nonproliferation Treaties

Nuclear weapons are far less dangerous if their owners cannot detonate them far from home. To help discourage proliferation by making delivery more difficult, the world's major industrial nations signed an agreement in 1987 called the Missile Technology Control Regime (MTCR). At the time, these were the only nations, with the exception of China, capable of designing and building long-range missiles. They agreed not to supply other countries with missiles that could fly more than 186 miles (300 km) or carry nuclear, chemical or biological warheads.

These countries still abide by the MTCR, but technology is passing them by. Less developed countries are learning to make their own missiles. They also modify less powerful missiles that they have bought from MTCR countries. Also, countries do not always agree on whether a given missile or technology has a nuclear application.[4] Although China never signed the MTCR, it promised to abide by its provisions. It has failed to do so, however, and it has sold missiles of nuclear potential to countries such as Iran and Iraq.[5] Pakistan has also refused to sign the agreement and has been selling missiles to countries that could soon have nuclear capability. Missile technology has been leaking out of the former Soviet Union, and missile technicians have emigrated from there in search of work in less developed countries. A group of thirty-six rocket engineers on their way to North Korea were stopped at the airport in Moscow. In 1995, fifteen Russian scientists were known to be sending information to Iran via computer modem.[6]

In 1971, seven NPT members agreed among themselves to restrict sales of certain nuclear equipment. Eventually thirty-one signed the agreement, and more are expected to join in the near future. Known as the Nuclear Suppliers Group, these are the countries that supply most of the world's nuclear and "dual-use" products. They will sell products with nuclear applications only to approved clients. Those clients have agreed to international safeguards and have promised not to sell the products to third parties. Adherence to the agreement is voluntary, unmonitored, and unenforced.

The Treaty of Tlatelolco went into effect in 1968. Under the agreement, Latin American countries will not acquire or possess nuclear weapons. No nuclear weapons may be stored or deployed anywhere in Latin America. In a later amendment to the treaty, nations that had nuclear weapons agreed not to threaten any Latin American countries with nuclear attack or to help any Latin American nation acquire nuclear weapons.

The Treaty of Rarotonga is an agreement among nations in the South Pacific to keep their region free of nuclear weapons, except those in transit on the military ships and aircraft. It was signed in 1986.

The Outer Space Treaty went into effect in 1967. It prohibited the installation of weapons of mass destruction in orbit around the earth, on the moon, or on any other celestial body.

The Seabed Arms Control Treaty, which went into force in 1972, prohibits the installation of any military installation, including systems to launch nuclear weapons, on the ocean floor.

In 1996, the United States and forty-six African nations signed the African Nuclear Weapons Free Zone Treaty, also known as the Treaty of Pelindaba. The agreement prohibits the research, development, manufacture, acquisition, testing, and stationing of nuclear explosive devices in any country that signed the treaty. The United States promised never to launch a nuclear attack against those nations. The treaty also prohibited the dumping of nuclear waste in those nations.

The breakup of the Soviet Union soon led to four treaties that were meant to control inadvertent proliferation in the post-Soviet nuclear powers. The Agreement between the United States and Russia Concerning the Safe and Secure Transportation, Storage and Destruction of Weapons and Prevention of Weapons Proliferation allowed the United States to assist Russia in the destruction of some of its nuclear and other weapons. It was signed and took effect in 1992.

A similar U.S.–Russian Agreement Concerning the Disposition of Highly Enriched Uranium Resulting from the Dismantlement of Nuclear Weapons in Russia allowed the conversion of HEU to LEU to be used in nuclear power plants. The agreement also provided for measures to prevent the proliferation of HEU. It was signed and became effective in 1993.

The Agreement between the United States and Belarus Concerning Emergency Response and the Prevention of Proliferation of Weapons of Mass Destruction took effect in 1992. Under the treaty, the United States agreed to help Belarus prevent proliferation of weapons of mass destruction and to respond to any emergencies relating to the removal of nuclear weapons from its territory. Two similar agreements, one between the United States and Ukraine and one between the United States and Kazakhstan, were signed in 1993.

The signing of the Limited Test Ban Treaty in Moscow, 1963

Progress toward a ban against the testing of nuclear devices took thirty-five years. The Limited Test Ban Treaty went into effect in 1963. It prohibited the testing of weapons in the atmosphere, under water, or in outer space. Indirectly, the treaty limits tests to underground explosions.

The Threshold Test Ban Treaty prohibited underground nuclear explosions in excess of 150 kilotons. The treaty was signed in 1974 but went into effect only in 1990.[7]

In 1995, President Clinton declared that the United States would not test any more nuclear weapons through actual detonation. (New technologies were making it possible to test the weapons without actually setting off a fission explosion.) In 1995, when France conducted what it promised would be its last tests, a worldwide uproar embarrassed the country though it failed to stop the test program. France indicated that it would sign a Comprehensive Test Ban Treaty (CTBT) after the tests were completed.

On September 24, 1996, most of the world's nations signed the CTBT. It outlawed the explosion of any nuclear device in the atmosphere, underground, or in outer space. The treaty itself will not inhibit nuclear proliferation, but it will remove the appearance of threat caused by nuclear tests.

Technical Problems

Disposing of nuclear weapons is no easier than building them. For that reason, the arms reduction agreements reached years ago will not be fulfilled for years. Russia is able to dispose of its 29,000 warheads at a rate of only about 2,000 per year. The United States, still not wanting to be too far down on the lower side of the balance of power, isn't moving much faster.

Some of the problems are technical. Destroying a bomb safely is a complicated matter.[8] First the weapon must be transported to the facility to be dismantled. The dangerous and potentially very valuable bomb travels in a special convoy of heavily armored vehicles that must be protected from everything from ambush to accident. In the United States, most bombs are taken to the Pantex complex near Amarillo, Texas.

The fissile "pit" of the warhead is not the only dangerous part of the bomb. It is surrounded by chemical (nonnuclear) explosives. If detonated, they would compress the core into critical mass. At best, the chemical explosion would scatter radioactive fuel over a wide area.

During the dismantling process, the pit and its explosive sphere are detached in a low, reinforced building under a concrete dome. The structure is designed to cave inward in the event of an explosion, dooming the technicians but trapping most of the radioactive fuel.

The Pantex plant in Amarillo, Texas, where most nuclear bombs in the United States are taken to be dismantled.

The nuclear fuel is placed in a special container for storage in a bunker designed to withstand earthquakes, fires, and just about any attack short of nuclear explosion. Storage in Russia is reputed to be less protected. An Associated Press report quoted a Russian as saying that "potatoes are better guarded" than nuclear materials, and reported that a thief stole HEU submarine fuel by simply cutting a padlock with a bolt cutter.⁹

At the Pantex plant, the most secret parts of the bomb mechanism are sealed in a virtually impenetrable shell. Although the computer components inside contain valuable gold, silver, platinum, palladium, copper and nickel, the whole mass must be destroyed. First the unit is frozen in liquid nitrogen to make it fragile; then it is hammered flat.

In the mid-1990s, Pantex was disassembling seven warheads per day during one shift of 400 to 500 workers, five days a week. The Department of Energy is planning to dismantle 2,000 bombs per year until the end of the decade.

Russia has been using four facilities to dismantle bombs at a rate equal to that of Pantex. If Russia plans to retain about 3,000 bombs, its stockpile should be dismantled by the middle of the first decade of the twenty-first century.

The most formidable problem is the final disposal of the nuclear fuel. Being of weapons-grade uranium or plutonium, the fuel is among the most radioactive substances on the planet. The half-life of plutonium-239 is 24,000 years. The half-life of uranium-235 is 700 million years. That means they will be dangerously radioactive for longer than humankind has been using fire. During that time, the fuel must be isolated from the environment. Any life form coming near it can be injured or killed. If scattered into the winds and waters of the world, it could cause cataclysmic damage.

Nuclear fuel being stored at a power plant in Oregon before it is removed for permanent disposal

Disposal of nuclear weapons fuel, therefore, must be more permanent than anything in the world, more permanent than bedrock. Geologists have found it impossible to calculate how the depths of the earth will shift over the next thousand centuries. Likewise, historians warn of the impossibility of predicting what civilization will be like a thousand years from now, let alone a thousand centuries from now. And they have no way to predict whether future technology will be able to dispose of or neutralize radioactive materials.

For the time being, to keep these materials inaccessible to terrorists and black marketeers, permanent disposal must be permanent indeed. It must not be within the reach of anyone now or a thousand years from now. Until such permanent disposal is devised, disposal must be temporary, that is, safe yet accessible for disposal later. Current storage in the United States, at Pantex and other nuclear facilities, is supposedly secure from anything except very heavy military attack. Still, almost 6,013 pounds (2,734 kg) of plutonium and 2,178 pounds (988 kg) of HEU cannot be accounted for at three U.S. nuclear defense facilities.[10]

In Russia, security is much looser. Furthermore, the Soviet government apparently did not keep accurate records of all plutonium production and disposal, so the Russian government is not sure how much plutonium and weapons-grade uranium it has, where it is, and whether or not any has been stolen. Russia is also hesitant to let the United States or the IAEA inspect its nuclear production, processing, and storage sites.

Technology may offer partial, temporary solutions to the storage problem. One is to mix the nuclear fuel with glass to form cylinders that can be buried or otherwise protected. They would still be highly radioactive, but it would be technically very difficult for anyone to separate the glass from the fuel to make a bomb. Unfortunately, the original mixing process is also technically difficult.

Another solution is to mix weapons-grade U-235 with industrial-grade U-238, which cannot be used for weapons. Again, the end product would still be dangerously radioactive, but it would be very difficult and expensive for an unauthorized user to purify the mixture into a fuel.

Scientists are also suggesting that plutonium and uranium be processed into plutonium oxides and uranium oxides, which can be mixed to form mixed oxide fuel (MOX). This fuel can be burned in nuclear power plants. This is commonly done in Europe. In the United States, however, the public worries about having concentrations of plutonium in power plants near cities. There is also political pressure resisting the movement of spent fuel to plutonium-processing plants and from there to power plants.

Russia, too, has technical and political obstacles to the disposal of plutonium. Technology is less advanced and the government has less money to resolve the problem. In fact, Russia does not see plutonium as a problem. Rather, it is seen as a national resource, something of value that had been produced at great cost. Having served an initial purpose as a fuel for nuclear bombs, it is now serviceable as fuel for power plants.[11] Even though it isn't very secure, they want to keep it. It's just too rare, valuable, and powerful to throw away.

Chapter 5

The Post-Soviet Nuclear Powers

When the Soviet Union collapsed and broke up in 1991, its economy was near chaos, its government was hopelessly corrupt, and various republics of the Union were declaring their independence.

Economically, the Soviet Union was virtually bankrupt. The communist economy, tightly controlled by the central Soviet government, had barely worked. When Mikhail Gorbachev started loosening government controls, the Soviet Union had no other economic system to replace it. Boris Yeltsin further reduced government control, and the economy spun out of control into deep depression.

The Soviet Union was also morally bankrupt. The government was hopelessly corrupt. Almost all social aspects of Soviet life were controlled by a megalithic bureaucracy. It was too cumbersome to work, and the bureaucrats often worked only for bribes. Beneath the appearance of a centrally controlled communist economy, a corrupt and uncontrolled form of free, albeit criminal, enterprise was working. As the communist system collapsed, this underground economy, which soon became an all-pervasive organized-crime syndicate, replaced it.[1]

The Soviet Union was also politically fractured. As the largest country in the world, it was actually a shaky union of many different republics and cultures. Several had been overrun by Russia before the Russian Revolution in 1917. Many were overrun later. Most of them disliked Soviet rule. When the central government weakened its control over them, they began to declare independence. At the same time, the Soviet "satellite" countries of Eastern Europe, which had been under Soviet domination since World War II, began to exercise independence.

By the end of 1991, the Soviet government was too weak to control its member republics. It was failing to control the economy and provide its citizens with basic necessities. The bureaucracy had simply stopped working. In December of that year, the Soviet government voted itself and the Soviet Union out of existence.

Fifteen countries emerged from the ruins of the Soviet empire. Four of them—Russia, Belarus, Ukraine, and Kazakhstan—inherited parts of the Soviet nuclear arsenal. They also inherited large stockpiles of radioactive materials.

Under ideal circumstances, these weapons and warhead fuels would have been put under tight security and dealt with in some organized manner. Unfortunately, the post-Soviet situation was far from ideal. The new republics had come into existence over the course of a few months or weeks. They had no experience at government. They had never owned nuclear arsenals before. Now they owned not only nuclear weapons, delivery systems, fuels and parts, but also the Soviet Union's most difficult endowments—bankruptcy, bureaucracy, corruption, and chaos. They had no money for food, medicine, or energy, and no system for exercising government control over virtually anything, including weaponry. Government officials, from presidents to police, were hopelessly corrupt, and sub-cultures within some republics were already beginning to demand independence.

It's hard to imagine a worse place to leave 30,000 nuclear weapons, 125 tons (114 mt) of plutonium and 1,000 tons (900 mt) of weapons-grade uranium. Since Russia was the dominant country to come out of the Soviet Union, it quickly pulled in all—or at least the world hopes all—the tactical nuclear weapons from Belarus, Ukraine, and Kazakhstan. These were the warheads that could have been used or transported relatively easily. The larger strategic weapons were protected by secret codes that would make them hard, but not impossible, to launch or detonate. Russia has been very secretive about the numbers, locations, and security devices of its nuclear weapons.

Belarus quickly decided that it did not want to be a nuclear power. It agreed to the provisions of START I and signed the NPT. All its strategic weapons were due to be shipped to Russia by the middle of 1996. Kazakhstan followed suit, transferring all its nuclear weapons to Russia by mid-1995.

Ukraine decided more slowly. It wanted to remain completely free of nuclear weapons and nuclear energy. Radioactive contamination from the accident at the Chernobyl atomic power plant in 1986 had left Ukrainians with a loathing of all things nuclear. Ukraine continued to operate the Chernobyl plant, however, because the country lacked other sources of energy. For three years after the collapse of the Soviet Union, it held onto its 1,650 nuclear weapons systems, using them as a bargaining chip. It wanted guarantees that Russia, Britain, and the United States would respect its territory and borders and would not use economic pressure against it. The real concern was with Russia. That giant neighbor had supplied Ukraine with oil and had been wanting to annex the part of Ukraine called Crimea.[2] The foreign powers agreed to respect Ukraine as

Ukraine soldiers preparing to destroy a ballistic missile, following their country's vow to become non-nuclear

a country, and in late 1994 the Ukrainian parliament voted to give up the weapons. They ratified the Nuclear Nonproliferation Treaty. The transfer of weapons, however, would take several years.

A Perfect Place for Proliferation

Authorities are concerned about several situations in the post-Soviet world. The most troublesome situation is that the Soviet government was never very accurate in its accounting of nuclear materials.[3] It started out by correctly registering how much plutonium and HEU was produced and distributed to its many research institutes, weapons laboratories and assembly plants, power plants, nuclear-waste storage facilities, and naval fuel depots. But after that original registration, each site was responsible for keeping track of its nuclear materials. There is no way of knowing whether these sites kept accurate records of plutonium that was lost in the course of processes, production, and transport. They had no systematized method of measuring stocks, and there was no national record-keeping. Central authorities had no way of knowing how much radioactive material existed, where it was, how much was missing, or where it may have gone.

As the economic woes of Russia and the other former Soviet republics increased, scientists and military personnel often went months without pay. Due to inflation, the value of their paychecks dropped by 70 percent or more. As much as a third of the staff at the Institute of Physics and Power Engineering, which stored enough uranium and plutonium for dozens of bombs, left their jobs, weakening security substantially. Those who remained at the institute and at other nuclear facilities could be tempted by anyone offering money for the ingredients of an atomic bomb. In a country where a few pennies bought a loaf of bread and monthly salaries were often under $100, the price of plutonium—thousands of dollars per gram—could be all but irresistible. As a director at the Institute told a New York Times reporter, "There are dishonest men in various levels, including in the bureaucracy. It's possible to buy anything in our country, including [nuclear] weapons and [radioactive] samples."[4]

In 1997, the Council for Foreign and Defense Policy, a respected Russian organization, warned that the Russian army might be close to collapse. Like workers across the country, soldiers and officers had gone months without pay. Russian Defense Minister Igor N. Rodionov warned that the country's vast nuclear arsenal, could become "uncontrollable."[5]

In September 1997, Alexander Lebed, a former Russian general and top security adviser to Russian President Boris Yeltsin, told CBS television program *60 Minutes* that ten nuclear bombs the size of suitcases were missing from the Russian arsenal. He had no idea what had happened to them. They could be armed by a single person within thirty minutes, he said, and each of the one-kiloton bombs could kill 100,000 people.

According to the editor of *Nuclear Fuel* magazine, in 1995 there were 950 sites in the former Soviet Union that had enriched uranium or plutonium. These sites were research institutes, weapons laboratories, assembly plants, power plants, and storage sites.[6]

With the general breakdown in the legal and police systems, the governments of the post-Soviet republics found it hard or even impossible to detect the theft of nuclear materials or to track down anyone who came to possess them. In 1994, several individuals were caught with plutonium and other radioactive materials in Europe, Russia and elsewhere (see chapter 8). Although Russia denied the possibility that the materials had come from its facilities, analysis indicated that they had. American intelligence agencies have also reported agents from Middle Eastern nations operating in post-Soviet republics in search of such materials. American professionals who have visited Russian nuclear plants have been asked, quite openly, whether they would like to buy radioactive materials.[7]

Authorities are also concerned with the very shaky political situation. Civil war threatens several post-Soviet republics, including Russia. Nuclear weapons and materials are by no means secure from attack by even a small army. They are tempting targets to any rebel group that might wish to quickly acquire a very deadly weapon or something that could be exchanged on the black market for conventional weapons.

The United States and Britain cooperated to secure one potential problem at a research reactor outside of Tbilisi, the capital of the former-Soviet nation of Georgia. Georgia is located in the Caucusus region, which has been politically unstable—torn by civil war and extensively controlled by organized crime. After two years of negotiation, it was agreed that U.S. Air Force planes would take 8.8 pounds (4 kg) of HEU and 1.76 pounds (3 kg) of highly radioactive waste to the Douinreay nuclear complex in Scotland.[8]

As discussed in chapter 8, the breakdown of law and order in these former Soviet countries has allowed powerful crime organizations to develop. Some fear that these organizations are more powerful than the governments themselves. With incredibly huge financial power, horrendously cruel threats, and connections with other international crime organizations, this "Russian mafia" is feared to have access to anything it wants, including atomic weapons, components, and fuel.[9]

The United States Tries to Help

Concerned for the safety of the world and especially itself, the United States has been offering considerable financial and technical aid to the post-Soviet republics. The most significant effort is known as the Nunn-Lugar program, so named because it was proposed by U.S. Senators Sam Nunn and Richard Lugar. It is also known as the Cooperative Threat Reduction Program. It is dedicated to helping Russia and the former Soviet republics transport, safeguard, and destroy their nuclear weapons and fissile materials.

Based on a 1996 arms reduction treaty, this intercontinental missile was one of the 130 to be destroyed in Ukraine.

As of the end of 1994, $1.2 billion had been allocated to help arms-reduction efforts in Russia, Ukraine, Belarus, and Kazakhstan. Most of that money is dedicated to helping transport and store warheads. Only $133 million was earmarked for the security of weapons-grade plutonium and uranium, and little of that was actually used to upgrade the physical protection or accountability of the materials.[10]

The Russian bureaucracy foiled at least one attempt at improvement. American officials tried to set up a $10 million system to monitor weapons-grade fuel, but the system ended up being installed at a low-enriched uranium plant that had no weapons fuel capability.[11] Russian officials said that the aid had been unnecessary for the purpose intended.

The corruption of the bureaucracies has made it hard to spend the allocated funds. These bureaucracies have a reputation for swallowing foreign aid without producing anything. The necessary accounting and safeguards have also hindered progress.

In November 1994, in a move that surprised the world, the United States airlifted 1,300 pounds (590 kg) of highly enriched uranium out of Kazakhstan. Much of the uranium was usable for weapons. The airlift was arranged after months of negotiations. American officials said that the fuel was very poorly protected and open to theft. Kazakhstan denied the poor security and said the fuel was never in danger. Fearing protests over its decision to bring such dangerous materials to U.S. soil, the U.S. government made no announcement of the operation until after it was over. The fuel was shipped to the federal government's Oak Ridge nuclear facility in Tennessee with the intent of later diluting it with low-enriched uranium for use in nuclear energy plants. Kazakhstan received a payment reportedly in the "low tens of millions of dollars" plus specialized nuclear equipment.[12]

In a move unthinkable just a few years in the past, the FBI opened an office in Moscow. Its two agents were given the task of dealing with the underground traffic in nuclear materials.

Russia insists that all of its nuclear weapons and nuclear materials are secure, but repeated reports indicate that much radioactive material has already disappeared from stockpiles and laboratory supplies. With Russia and many of its post-Soviet neighbors in chaos, and with many years yet to go before the thousands of weapons have been destroyed and the tons of radioactive materials are put under strong security, the situation can be called, at best, unpredictable. Unfortunately, there are many predictions, and few of them are good.

Chapter 6

The Status of Proliferation

Countries decide to "go nuclear" for several reasons. In some cases, they believe that an atomic arsenal will guarantee a certain level of national security. The threat behind a single bomb can hold at bay a country with much larger armed forces or even a much larger nuclear arsenal. Countries that see themselves as uncomfortably close to enemies, especially a lot of enemies or an especially powerful enemy, are the most likely to seek atomic military capability.[1]

Atomic capability can also be a powerful "bargaining chip" in negotiations toward political, military, or economic advantage. A country may exchange its nuclear capability for economic aid or conventional military assistance. Countries with serious economic problems are the most likely to develop an atomic bargaining chip.[2]

Some countries think that atomic weapons will give them the power they need to expand their territory or military influence. A country that believes itself in possession of some great religious or political truth may be tempted to give its beliefs nuclear strength. These countries tend to be the same ones that support terrorism as a political tool, and an atomic bomb can be a most tempting terrorist weapon.[3]

Countries may also be tempted by the status of membership in the nuclear club. In the international community, countries with atomic weapons tend to receive more attention and respect. In international negotiations, a country that has one of the world's most powerful weapons will feel less pressure to knuckle under to the persuasions of more powerful countries. Former French President Charles de Gaulle was not wrong when he said, "No country without an atom bomb could properly consider itself independent."[4]

In most cases of countries working toward a nuclear arsenal, motives are mixed. Indeed, it is easy for a country to see several advantages, and few disadvantages, to acquiring a nuclear bomb or two.

Until recently, it was assumed that a country would need ten years to develop the technology necessary to build a bomb. That long period gave the world plenty of time to apply diplomatic and economic pressure to end the development process. Recently, however, that development period has shrunk. Technologies have simplified. Specialized and dual-use tools are more widely available. Nuclear fuel is more available. A black market for fuel, weapons, and delivery systems either exists or soon will. It is increasingly possible for a country to come very close to assembling a bomb before anyone suspects a thing. At that point, diplomacy becomes less effective and the situation becomes more dangerous.

No single pattern explains how the various nuclear powers, seekers of nuclear power, and former nuclear powers have behaved. India progressed secretly and then surprised the world with a nuclear explosion. Israel developed atomic weapons and then used vague claims to deny their existence. Brazil and others began developing the technology but then stopped. South Africa developed the bomb, assembled a small arsenal, then renounced its nuclear power and destroyed its arsenal. Japan has made no overt moves toward developing a bomb, but Japanese politicians

Brazilian President Collor de Mello revealed his country's nuclear program in 1990 and promptly shut it down.

have admitted that Japan has the know-how, and they have suggested that an atomic arsenal might become necessary.[5]

A separate class of countries, often called "rogue regimes" for their disregard of international law and order, are suspected to be working on atomic technology. They are not likely to admit their intentions until they are ready to make use of their weapon. They are considered to be seeking atomic power for reasons other than defense. These rogue regimes are discussed in the following chapter.

Below is a list of some of the countries that have worked on atomic technology. Some have produced a bomb. Others may still be working on it. Others have stopped. Others may start.

Brazil

Brazil was not a party to the NPT when it began a secret nuclear weapons program in 1980s. Though a democracy, its military was very influential in the government and proceeded to upgrade uranium to a quality suitable for weapons. A bomb test site was built in the Amazon region. In 1990, shortly after being elected, then-president Fernando Collor de Mello revealed the existence of the nuclear program, terminated it, and closed the test site.

Without signing the NPT, Brazil and its neighbor, Argentina, agreed not to develop nuclear weapons and to have their nuclear facilities inspected by the IAEA. There are questions about how accurately the IAEA will be able to detect activities that took place five to ten years earlier. It may not be possible to determine if Brazil ever produced enough HEU to build a bomb. Brazil is known to have bought enriched uranium from China, and it may be impossible to determine definitely how much, or what became of it.[6]

Argentina

Like Brazil, Argentina was under military rule or influence during the 1980s, and during that time the nation built uranium enrichment facilities. Because Argentina had not signed the NPT, the facilities were not open to IAEA inspection. Imports of enriched uranium from China could have been used for a weapon but may have been re-exported to Argentinian-made nuclear reactors in Peru and Algeria. To prevent a nuclear arms race in South America, these two nations have agreed to inspections by the IAEA and have vowed not to develop weapons.[7] Argentina signed the NPT in 1995.

India

To the surprise of the whole world, India exploded a nuclear device in 1974. For the next twenty-four years, it denied having a nuclear arsenal yet refused to sign the NPT. Reports indicated that India could have assembled twenty-five bombs in a matter of days.[8] In May 1998, India again surprised the world again, this time with five underground nuclear tests. Almost immediately, Pakistan announced that it would detonate weapons in response.[9]

Almost all of India's international border is with Pakistan and China, both nuclear powers. Pakistan has been considered an enemy since both nations became independent in 1947, and one border skirmish came close to the nuclear threshold (see the section on Pakistan that follows).

Critics also worry that India may not be sophisticated and organized enough to manage a hostile nuclear crisis. The government has been a reasonably stable democracy since its independence, but there are signs of serious turmoil and lack of control. Several leaders have been assassinated. Areas near Pakistan have long been at the brink of civil war. Hindu-Muslim religious conflicts have often become widely violent. Neighboring Pakistan considers itself at a nuclear standoff.

When India conducted its nuclear tests in 1998, it became subject to the sanctions of the 1994 Nuclear Proliferation Prevention Act. The law ordered President Clinton to cut off almost all government aid to India, bar American banks from lending money to the Indian government, and stop exports of U.S. products with military uses. The law also required the United States to oppose aid to India by the World Bank and the International Monetary Fund. At the time, India was the world's largest borrower from the World Bank, with more than $40 billion in debts.[10]

Pakistan

Pakistan never signed the NPT, and until 1998, it denied that it had a nuclear capability. But when India detonated five bombs in the middle of May that year, Pakistan responded with an equal number by the end of the month.[11]

The common motivations for developing nuclear weapons certainly apply to Pakistan. It shares borders with two much more powerful and populous nuclear powers: China and India. India has been considered an enemy since 1947, when Pakistan broke away from India to form a Muslim nation. Since then, war has broken out between them three times, and the Indian territory of Kashir is under both Pakistan and Indian military control. In May 1990, when India held massive military maneuvers near the Pakistani border, Pakistan, fearing imminent invasion, may have transported a nuclear weapon to a fighter plane which was to have dropped the bomb on New Delhi. U.S. diplomacy helped separate the enemies in time.[12]

Pakistani Prime Minister Benazir Bhutto has never been allowed to visit her country's nuclear plant.

The United States was of some help in Pakistan's nuclear program. The United States needed Pakistan's cooperation in the resistance to the 1979 Soviet invasion of Afghanistan, which lies along Pakistan's northwest border. To retain Pakistan's cooperation, the United States contributed substantial economic and military aid. Although it was clear that Pakistan was seeking nuclear technology that could be applied to a bomb, the U.S. government took no action beyond mild protest.

It has long been estimated that Pakistan had produced enough highly enriched uranium for more than six nuclear devices, and perhaps could assemble as many as fifteen bombs almost immediately.[13] U.S. intelligence sources detected clear efforts by Pakistan to acquire the designs and components of a bomb. The Central Intelligence Agency said Pakistan had already built several bombs and bought nuclear-weapons technology from China. In 1990, therefore, President George Bush said that he could not certify that Pakistan did not have a "nuclear explosive device." Without that certification, Pakistan could no longer receive economic or military aid.

Nonetheless, in 1996 the Clinton administration agreed to deliver $368 million worth of weaponry. *The New York Times* quoted Republican Senator Larry Pressler of South Dakota saying, "We don't have a non-proliferation policy anymore. We have an arms bazaar." According to legislative aides who were briefed on the sale, the Clinton administration said that Pakistan would be a better ally if the weapons were delivered.[14]

There is much concern over Pakistan's ability to "manage" nuclear power. Its government is extensively corrupt. No prime minister has ever been allowed to visit Pakistan's Kahuta nuclear facility. Prime Minister Benazir Bhutto apparently had no say in the 1990 decision to escalate the border conflict with India to a nuclear level.

The Pakistani people, as a whole, appear to be proud to consider their country a nuclear power. The director of Pakistan's intelligence agency has been quoted as saying, "Nuclear capability is one symbol of Pakistan's sovereignty . . . . It is a symbol of national honor. This is one issue where if we compromise we will be dishonored as a people."[15]

Israel
Israel has not signed the NPT and has neither denied nor confirmed accusations that it has a nuclear arsenal of as many as 100 to 300 atomic weapons.[16] It probably received technical assistance from France and possibly received plu-

tonium from the United States. It has reason to perceive a need for nuclear power as a deterrent because it has been surrounded by enemies since its creation in 1947. Israel has a technological infrastructure adequate for producing atomic weapons and as much as 110–200 pounds (50–90 kg) of plutonium per year. It also has missile systems capable of delivering nuclear warheads throughout the Middle East. Israel may have cooperated with South Africa in a test detonation in the Indian Ocean in 1979, and it has been accused of stealing enriched uranium and crucial nuclear equipment from companies in the United States.

In the months before the NPT Extension Conference of 1995, Israel's refusal to sign the NPT and to permit inspections became a thorn in the NPT negotiations. While not admitting to possession of nuclear weapons, Israel argued that it feared nuclear attack from Iran, Iraq and Libya. Egypt, however, argued that if Israeli extremists came into power, they could threaten their Arab neighbors.[17]

Japan

Since the atomic destruction of Hiroshima and Nagasaki, Japan has been resolutely against the presence of nuclear weapons in its territory. (U.S. warships visiting Japanese ports are an unofficial exception.) When North Korea seemed to be building a bomb and was experimenting with a missile that could reach Japan, some Japanese officials were seriously suggesting that Japan should balance the threat with a nuclear weapon of its own. They claimed that, with Japan's advanced nuclear and space technologies, it was capable of quickly developing weapons and delivery systems. Later, Prime Minister Morohiro Hosokawa formally renounced the possibility and committed Japan to an extension of the NPT.

Japan has a major plutonium production plant, called Rokkasho, that separates plutonium from spent nuclear energy fuel. The plutonium is to be used as a fuel. Within thirty years Japan will have acquired twice as much plutonium as the United States and Russia had in the early 1990s. Already Japan is unable to account for all the plutonium it is supposed to have. More than 150 pounds (68 kg) of plutonium has disappeared from the Rokkasho facility—enough to fuel a dozen bombs the size of the one that destroyed Nagasaki.[18] Japan claims that this is normal, an inevitable result of plutonium getting "hung up" in pipes, gloves, and equipment. Some experts, however, say that the Japanese facility is too sophisticated to lose that much of the fuel.

It is feared that Japan could use its huge stockpile of plutonium, plus its renowned technical ability, to quickly build a massive nuclear arsenal.[19] The mere potential has tended to increase distrust in the region.[20]

South Korea has already told the United States that if Japan has permission to extract plutonium from spent fuel from the United States, South Korea wants the same. North Korea uses the Japanese program to justify its own.[21] The situation could turn into a plutonium race. If the international political situation turns tense, an arms race could develop and escalate quickly. If other countries in the region fear that Japan has a nuclear arsenal, they will want to develop an arsenal, too. Japan, then, might feel obliged to keep up.[22]

South Africa

South Africa is known to have had a uranium enrichment plant and could have produced enough fuel for twenty to thirty bombs. The motivation would have been the general

African and national resistance to South Africa's rigid racial separation policy known as apartheid. In 1991, however, as South Africa was dismantling its apartheid legal system and preparing for possible rule by blacks, it signed the NPT after becoming the first nation to voluntarily give up a nuclear arsenal. Authorities have not been able to definitely conclude how much fuel and how many weapons South Africa produced—its government says six—and whether all the fuel has been accounted for. It has been alleged that South Africa may have exported some of its fuel to Israel. The facilities that produced the first bombs and components are presumably still in existence, making it relatively easy for South Africa to rebuild its arsenal. That capacity is especially worrisome since South Africa is building a missile, the Arniston, which can carry a nuclear warhead.[23]

Taiwan

Taiwan is considered a potential new nuclear power because it is next to a much larger power, China, which has nuclear weapons and has long threatened Taiwan's existence. Twice, in the early 1970s and late 1980s, Taiwan secretly began trying to produce plutonium for use in nuclear weapons. In both cases, U.S. diplomatic efforts brought the program to a halt. Today, Taiwan has a very successful nuclear power program that could be adapted to produce bomb fuel. Development of a bomb would probably take fewer than the ten years that less developed countries might need. If China became more aggressively threatening, Taiwan might be tempted to quickly acquire nuclear power, leaving the world with little choice but to accept its membership into the nuclear club.[24]

The Proliferation of Plutonium

By the year 2010, 550 tons (498 mt) of plutonium will be separated from the spent fuel of nuclear power reactors.[25] Nothing can be done to destroy it. Virtually all of it will continue to exist for thousands of years. At best it can be mixed with other materials, making it inaccessible. Considering how little is needed to build a bomb—12 pounds (5.4 kg) would be plenty—or to act as a deadly pollutant, that quantity is truly disturbing. A few pounds missing from 550 tons would be the equivalent of a word missing from this book. The loss might be equally hard to detect.

The proliferation of plutonium could lead to serious problems. Plutonium has become a fuel commodity that is legally bought and sold on the international market. The NPT promotes its use as an energy fuel. Low enriched uranium is good enough to use as a fuel and currently costs one-quarter to one-eighth as much as plutonium fuel. Nonetheless, many countries are considering the use of weapons-grade plutonium as an energy fuel. One reason for the use of this inherently dangerous fuel is that it's a by-product of the decay of low-enriched uranium fuel. Safety and economic considerations aside, it makes sense to use the plutonium as fuel rather than go to the expense of disposing of it. Disposal is difficult because plutonium is so radioactive and has such a long half-life.

The potential problem with increased inventories of plutonium and its movement as a commercial product is that some of it might leak, get lost, or find its way to the wrong hands. Even the closest IAEA inspections cannot account for all plutonium. The agency admits that it could miss the diversion of 600 pounds (270 kg) of plutonium per year worldwide.[26] It could never been known how much got

Disposing of radioactive nuclear waste is difficult but very important. Improper disposal can lead to contamination and disease.

"hung up" in pipes and how much might have been diverted for secret purposes.

When Britain was preparing to test a new plutonium processing plant, U.S. Representative Pete Stark of California said that the plant was "a direct threat to international security, bringing an additional 59 tons [53.5 mt] of plutonium into circulation over the next ten years." He and other legislators signed a letter to President Clinton urging him to pursue a global ban on the production of all fissile material that could be used to make nuclear weapons.[27]

Plutonium may or may not be a safe and cost-effective energy fuel. Its acceptance as a commercial product, however, can only complicate the task of keeping new weapons out of new hands. To prevent such proliferation, the IAEA will have to greatly improve and expand its role as inspector, and international treaties will have to deal with the problem more directly.

Chapter 7

The Rogue Regimes

The world is especially worried by nuclear proliferation among the so-called "rogue regimes." These countries use words and action to threaten other countries, support terrorism, or flaunt international treaties such as the NPT and decisions reached by the United Nations. It is assumed that they want nuclear weapons for attack rather than for deterrence. Unfortunately, the rogue regimes are among the countries most actively and secretly trying to develop nuclear weapons.

Rogue regimes known to have—or have had—programs dedicated to developing nuclear weapons include North Korea, Iraq, and Iran. Libya apparently has no such program, but its leader, Muammar Quadaffi, has publicly announced that he is in the market for an atomic weapon.[1] Algeria has a slowly developing nuclear program that may be working toward a weapons capability.

The tendency for nations to need ten years to develop a weapon does not apply to these nations. Often these countries cooperate with other countries to acquire more quickly the technology, equipment, components, and ingredients needed to produce a weapon. Sometimes these countries are other rogue regimes or other authoritarian countries with similar religious or political beliefs. Often they are the countries most opposed to nuclear proliferation, the highly developed democracies such as the United States and Germany.[2] Even if the countries need ten or more years, since their programs are secret, no one knows of their progress until their later, advanced stages.

China deserves special mention here. It is not included in the list below because it is not a proliferator, that is, a country recently seeking a nuclear weapon. China has made many friends among the rogue regimes, however, and its sales of missiles and nuclear technology could qualify it as a rogue supplier. Despite the NPT and international treaties preventing the proliferation of arms, China has sold nuclear and missile technologies wherever it has found a decent price. Thus Iraq has missiles that can reach Israel, North Korea has missiles that can reach Japan, and Iran has missiles that can reach the oil fields throughout the Middle East region. Argentina and Brazil have bought enriched uranium, Iran and Algeria have bought small research reactors, and Iraq has bought restricted equipment not available from more conscientious countries.[3]

Libyan leader Muammar Quadaffi has made it known that his country would purchase nuclear weapons.

China may have provided Pakistan with a nuclear weapon design and enough uranium for one or two bombs.[4]

Both Iraq and North Korea surprised the world with unexpected progress in their nuclear programs. Authorities who monitor proliferation activities and programs term the phenomenon *unexpected acceleration*. Unexpected acceleration is sudden progress toward a weapon, an advancement of technology that was not foreseen. Unexpected acceleration does not allow peaceful diplomatic efforts to become effective. Such efforts take too much time. In the cases of Iraq and North Korea, the United States and the United Nations had to take drastic action to curtail the programs.

The following discussion reveals the history and status of the nuclear activities of rogue regimes.

Iraq

Under the tyrannical leadership of Saddam Hussein, Iraq has been pursuing nuclear power for more than twenty years. Israel, fearing for its survival, attacked and destroyed Iraq's Osirak nuclear research reactor in 1981 before it was finished. The attack, however, was only a temporary setback. Iraq, like other rogue regimes, learned that if it was to develop an atomic weapon, it would have to do so in secrecy. Iraq's nuclear program went into hiding. Iraqi agents in foreign countries secretly—and sometimes openly—bought the equipment needed to process and purify uranium. Iraq was able to buy enough parts and equipment to build a sophisticated electromagnetic isotope separation process. Some of the parts were easy to buy because the technology was considered obsolete.[5] Iraq was just beginning to enrich uranium when allied forces attacked Iraq in the Gulf War that followed Iraq's invasion of

Saddam Hussein, Iraq's leader, has been pursuing nuclear power for many years.

Kuwait. Many of the bombing raids were specifically aimed at Iraq's nuclear infrastructure.

The negotiated end to the war gave the IAEA unrestricted access to all Iraqi nuclear facilities and information related to weapons of mass destruction. Weapons of mass destruction include chemical and biological weapons. Iraq also had to give up all ballistic missiles with a range of more than 93 miles (150 km) and any equipment, materials, components, and data related to atomic weapons.

Iraq resisted each step by the IAEA and may have succeeded in hiding considerable information, equipment, and nuclear fuel. In one case, an inspection team was held up at the front gate of a military base while equipment was hurried out the back gate. By 1998, the IAEA was still searching for information.

The IAEA inspection teams found that many of the bombing raids had destroyed empty buildings. Many elements of Iraq's weapons of mass destruction had been hidden elsewhere. The teams also analyzed equipment to determine Iraq's nuclear suppliers. The countries whose companies contributed the most were, in order, Germany, Switzerland, Britain, France, and the United States.[6] Iraq refused to release the names of all its suppliers, and the IAEA refused to release the names of all the suppliers it had detected.[7]

The IAEA was surprised to find out that Iraq was perhaps only about three years away from completing a weapon.[8] Iraq's success in hiding such a large, multi-facility production intensified international discussion about the effectiveness of IAEA safeguards. IAEA Director General Hans Blix told his board of directors that the very essence of nonproliferation commitments and their verifications is that neighbors, regions, and the world at large should be able to rely upon them with confidence and without risking disastrous surprises.[9]

In 1992, an Associated Press article reported that fifty Russian nuclear scientists were working in Iraq.[10] By late 1994, despite the trade restrictions, Iraq had already become the fourth-largest military power in the world, with 80 percent of its military complex rebuilt.[11]

In 1998, Iraq protested the presence of too many Americans in the UN inspection team. For several weeks, the United States and Iraq bordered on war. During that time, inspections were prohibited. In April, just after the IAEA said it had found no indications of prohibited nuclear materials, other experts warned that the situation was still dangerous. David Albright, president of the Institute for Science and International Security, said that Iraq could still build a bomb in less than a year; all it lacked was the fuel, which, he said, was available on the black market. Paul Levanthal, president of the Nuclear Control Institute, said the IAEA conclusions "should be viewed with disbelief . . . the agency has been consistently wrong on the subject of Iraq. You have to assume that a nuclear weapon remains Saddam's ultimate prize. It is also [wise] to assume that there is a small workable weapons project in Iraq." He said that the world could assume that bomb-making materials were available on the black market.[12]

Iran

Iran has been seeking an "Islamic bomb" since 1979, but political turmoil and a long war against Iraq hindered progress. In the 1990s, however, Iran intensified its efforts to develop a bomb by importing know-how and hardware.

Observers did not foresee an Iranian bomb until the first decade of the twenty-first century. In early 1995, however, a U.S. official in Tehran said that Iran was less than five years from its first bomb. Progress depended mostly on the building of two large reactors in the Persian Gulf city of Bushehr. Iran had also contracted the construction of two Russian and two Chinese reactors. U.S. Secretary of State Warren Christopher later said that an Iranian bomb was at least fifteen years away as long as it did not acquire fuel from a foreign source.[13] In May 1995, he also said, "Based on a wide variety of data, we know that since the mid-1980s, Iran has had an organized structure dedicated to acquiring and developing nuclear weapons," and that Iran's nuclear program was "pursuing the classic route to nuclear weapons which has been followed by almost all states that have recently sought a nuclear capacity."[14]

A few months later, Christopher went to Russia to try to persuade the government not to go ahead with plans to help Iran build two light-water reactors. Light-water reactors are especially useful for producing plutonium. One of these reactors could produce hundreds of kilos of plutonium per year.[15] He revealed a secret report that showed Iran had been trying to buy fissile materials from former Soviet republics and was clearly in the market for equipment that could be used only to build an atomic weapon. When Russian officials said they planned to go ahead with plans to help build the plants, Christopher said, "Russia will rue the day it cooperated with the terrorist state of Iran. . . ."[16]

U.S. diplomatic efforts were more successful with the reactors which China was to build. China agreed to cancel the contracts. The project never went beyond the initial designs.

According to a 1995 *New York Times* article, senior Israeli officials have stated that an Iranian bomb is currently their country's biggest security concern. If Iran nears completion of its new reactors at Bushehr, the officials said that Israel would consider a bombing raid like the one that destroyed Iraq's Osirak reactor. The Israelis also suspected that Iran and Pakistan had signed a secret agreement to work together on atomic weapons.[17]

Another *New York Times* article in 1995 identified a trail of smuggling operations from a small airport in

Germany to Iran. The airport was owned by three Iranians who are reputed to be among the biggest arms dealers in Europe. Nuclear equipment was broken down into unidentifiable parts, loaded onto small planes at night, and flown to other European cities. The smuggling routes, which were also used to smuggle drugs into Europe, were the same as those Iraq and Pakistan had used to acquire nuclear equipment. German intelligence officials said that despite knowing of the activities, they could at best only make them slower and more difficult.[18]

Meanwhile, Iran has been acquiring North Korean Scud-C and seeking to acquire Nodong I missiles, and China is building an assembly line in Iran to produce intermediate-range missiles.

An Iranian official denied any such nuclear military program, stating that "the IAEA [has said] that our research is 100 percent civilian and has no military purpose. We are for a Persian Gulf area that is free of nuclear weapons."[19]

North Korea

North Korea, a hard-line communist country, has engaged in very sly deception to develop an atomic capability, to avoid the requirements of the NPT, and to wangle advantage over the United States and South Korea.

North Korea began a nuclear program in the mid-1960s and a weapons program in the early 1980s. In 1985 the country signed the NPT. The first inspections were scheduled for eighteen months later. But just as the deadline approached, North Korea claimed that it had received the wrong forms that were to be filled out beforehand. The deadline was extended by eighteen months, but then North Korea missed that deadline as well. In 1989, North Korea secretly shut down the plant and probably extracted enough plutonium for one or two atomic bombs.

In 1990, North Korea said it would withdraw from the NPT if the United States did not remove all nuclear weapons from South Korea. In 1991, North and South Korea signed an agreement banning all nuclear weapons from the Korean Peninsula. In January 1992, North Korea agreed to IAEA inspections. In 1993, however, it denied inspections of two secret nuclear fuel reprocessing sites, which were believed to be producing plutonium. When the IAEA demanded a "special inspection," that is, an unrequested inspection, North Korea announced that it was withdrawing from the NPT. In June it changed its mind but still prevented the inspections. By the end of the year, the IAEA stated that it was unable to assure the world that North Korea had not diverted fuel for weapons production. In May, 1994, North Korea defueled a reactor without waiting for IAEA inspectors. It would have been a good time to secretly divert fuel to a purification process, though IAEA inspectors believe they accounted for all of it.

Meanwhile, North Korea accelerated construction of two much larger reactors. One, scheduled for completion in 1995, would be able to produce fuel for a dozen bombs per year. The other, to be completed in 1998, could produce thirty to forty bombs per year.

Negotiations with the United States involved preparations for war on both sides of the Korean border in 1995. Negotiations led to North Korea agreeing to adhere to the NPT, to allow inspections, to dismantle its two uncompleted reactors, and to freeze its nuclear weapons program. In return, it would receive two new nuclear reactors worth $4 billion, paid for by the United States, Japan, South Korea, and other countries. These reactors would be less appropriate for the production of plutonium. The United States would also give North Korea oil until its new reactors could provide energy.

The agreement, which was reached by the Clinton administration, was criticized by Republican Senate Majority Leader Robert Dole and others. North Korea was being allowed to deny inspections of nuclear-waste sites where it would have been possible to detect whether plutonium had been produced. North Korea was not required to hand over any bombs it may have already produced. North Korea countered by calling Senator Robert Dole "a deplorable political ignoramus" for his attempt to block the agreement, which North Korea felt was "flawless."[20]

The agreement was also criticized as encouraging other countries to develop nuclear capability and to disregard the NPT. By so doing, North Korea had reaped a nice profit and most likely retained enough plutonium for a bomb or two. During the next few months, North Korea again hesitated and found several reasons not to go ahead with the original plan.

In the middle of 1998, American intelligence agencies detected construction of an elaborate underground nuclear complex that could produce half a dozen bombs within two to five years. Construction of the new nuclear plants had not yet started, and North Korea was complaining that the United States had fallen behind on its promised fuel oil deliveries.[21]

The world's experience with these three blatant nuclear proliferators demonstrates the difficulty of enforcing nonproliferation. They demonstrate that it is quite possible for a country to ignore the NPT, prevent IAEA inspections, develop nuclear capability secretly, move faster than estimated, and continue nuclear programs even after military or diplomatic countermeasures have been taken. Ultimately, the success of these rogue regimes may erode confidence in the NPT and the IAEA and encourage other nations to build atomic arsenals to deter nations within the range of missiles.

Chapter 8

Deadly Traffic

The Soviet Union collapsed and broke up very suddenly, leaving a vast empire without an effective government or legal system. The consequent power vacuum gave birth to an explosion of organized crime. The "Russian mafia" was a natural development of the corrupt bureaucracy that had reached into every town, business, and government agency in the former Soviet empire. This mafia was not a single organization but a loosely connected network of more than 5,000 individual gangs; some 200 of them were large and well organized. About 100 of them had relations with crime groups in other countries.[1] The Russian mafia reached into every major capital city in the world through the former Soviet intelligence agency, the KGB. Almost instantly, the Russian mafia had branch offices throughout the world and vast data banks with information on millions of individuals, companies and governments.[2]

With the Russian government powerless to stop or limit it—generally, police and military officials were and are still part of the organization—the Russian mafia amassed unbelievable wealth. Intimately involved with the Soviet and post-Soviet military, it stole and sold entire arsenals of old Soviet weaponry, including machine guns, surface-to-air missiles, and tanks. When Russian soldiers pulled out of East Germany, 81 tons (73 mt) of ammunition were discovered to be missing.[3] Russian generals have authorized military planes to transport weapons and ammunition to private buyers. Illicit nuclear materials were flown into former Soviet bases in East Germany.

The mafia bribed factory directors to hand over entire production lots that were then exported and sold. Entire tankers of oil and trains of ore, metal, and timber were exported, with tens of billions of dollars going into the pockets of racketeers.[4] In 1994, President Boris Yeltsin reported to the Russian Parliament that organized crime was the biggest threat to the security of Russia.

The mafia soon controlled most of Russia's two thousand banks, making it possible to launder huge amounts of money and to succeed in several international credit scams. Perhaps as much as 30 to 40 percent of Russia's gross national product is controlled by the mafia.[5] The amount of cash flown from New York to Moscow between 1994 and 1996, hundreds of tons of $100 bills, exceeded the value of all the Russian rubles in circulation. Law-enforcement officials assumed it was "laundered" money going to mafia-controlled banks.[6]

As the network of crime spread across the former Soviet republics, a smuggling route developed across Eurasia. Its flow carried goods to Europe from as far away as Mongolia and Afghanistan. A huge bazaar of smuggled goods, including weaponry, drugs, and nuclear materials, grew in the Nowy Targ Valley in the Tatra Mountains of Poland. On some weekends, tens of thousands of buyers from dozens of countries shop there for everything from socks stolen from Russian factories to hashish from Pakistan to machine guns stolen from military bases to nuclear materials stolen from power plants.[7]

The Russian mafia has held summit meetings with organized crime figures from Colombia, Japan, China, Italy, and the United States. These organizations cooperate to engage in virtually every illicit business in the world. They "launder" billions of dollars by passing money through different banks, exchanging currencies, and buying and selling products to hide the sources and the destinations of their huge revenues. Their profits have exceeded $1 trillion, making them wealthier than all companies and most governments. Each year, an estimated $500 billion is laundered to hide its origins. Organized crime is now the fastest-growing industry in the world.[8]

The Mafia Goes Nuclear

The real terror of such powerful and wealthy global organized crime became apparent when nuclear materials began appearing in Europe. Police reported 4 cases of attempts to buy radioactive materials in 1990, 41 cases in 1991, 158 cases in 1992, and 241 cases in 1993. Not all the cases really came to involve radioactive materials—more

In 1994 FBI Director Louis Freeh warned Congress that the Russian mafia is interested in nuclear weapons purchases.

than half of the cases were scams—but it can be assumed that many such crimes went undetected. For good reason, in 1994 FBI Director Louis Freeh told the U.S. Congress that Russian crime groups "may already have the capability to steal nuclear weapons, nuclear weapons components, or weapons-grade nuclear materials."[9]

A handful of cases revealed how real and serious the problem was. In May 1994, German police raided a house in search of counterfeit money but found 0.2 ounces (6 gm) of very pure weapons-grade plutonium. Although far less than the amount needed for a bomb, its presence in somebody's garage clearly indicated that one of the most controlled substances in the world had not been controlled enough. It was the first time plutonium had been found not under the control of a government or responsible organization.[10]

In early August of that same year, an undercover German police operation seized 10.5 ounces (300 gm) of plutonium and uranium being smuggled into Munich by three Colombians on a plane arriving from Moscow. The Russian minister of energy happened to be on the same plane. At first Russian officials denied the possibility that weapons-grade fuel had been removed from Russian facilities. Later, however, Russian atomic-energy officials identified three possible sites where the fuel could have come from.[11] German intelligence revealed that a few weeks earlier Iranian agents had been in Germany trying to buy ingredients and equipment that could be used to build an atomic bomb.[12]

Later that month, Hungarian police arrested someone trying to sell 4.4 pounds (2 kg) of mixed radioactive material from atomic power plant fuel rods. The asking price was $40,000.[13]

At the end of August, Estonian authorities arrested three Russians and an Estonian with nearly 7 pounds (3 kg) of low-enriched uranium. Moscow authorities admitted that the uranium had been stolen from a Russian military plant. Estonian police determined that Estonia was only a transit point on a route headed toward Europe. Russian authorities arrested two young men who had stolen 21 pounds (9.5 kg) of industrial-grade uranium-238 from a nuclear weapons complex. The thieves had hoped to trade the deadly material for a video camera and cassette recorder.[14]

In December 1994, police in Prague found 6 pounds (2.7 kg) of highly enriched uranium in the back seat of a Saab. Three nuclear workers from the Czech Republic, Belarus, and Ukraine had been offering to sell the uranium for "several million dollars." Analysis determined that the HEU could have come from any of 950 sites in the former Soviet Union.[15]

In 1995, the CBS television program *60 Minutes* and *U.S. News and World Report* simultaneously broke a story about 4 tons (3.5 mt) of beryllium seized in Lithuania in 1993. The metal is used to increase the power of atomic weapons by reflecting neutrons into the core as the explosion begins. It is also used in missile-guidance systems and high-performance aircraft. Twenty-seven crates of the silvery metal cubes had apparently come from a nuclear research station and the theft had undoubtedly involved the Russian mafia. According to documents, the shipment had included 20 pounds (9 kg) of radioactive cesium valued at $100,000, but the cesium had disappeared before authorities found the beryllium. An unknown buyer in Switzerland was offering $24 million for the beryllium, ten times more than the market price. The news reports indicated North Korea as the most likely destination. Authorities also learned that the nuclear research station had sent out an earlier shipment of beryllium, but it has never been found.[16]

As the press reported these and other incidents, a British nuclear consultant reported that on a recent trip to Russia, several managers at nuclear-power plants had asked him if he was interested in buying radioactive materials. Apparently, since atomic bombs were no longer being built, plutonium stockpiles were building up. People in control of them thought that selling plutonium was a perfectly legitimate way of disposing of it. "The Russian nuclear industry is leaking like a tea bag," the consultant said.[17]

A nuclear-weapons expert with Greenpeace, an international environmental-action group, was offered an atomic weapon for $250,000. He was taken onto a Russian military base in Germany and shown how locks, codes, and alarm systems could be bypassed. The base was under the guard of only twelve officers, and the plan was to simply drive a truck up to the storage building and load the bomb onto it. A coup attempt in Moscow foiled the plan.[18]

The Bulletin of the Atomic Scientists reported on an incident that demonstrated how easily nuclear materials can be stolen. The article reported that, in 1993, a Russian military captain slipped through an unprotected gate at a shipyard, climbed through a hole in a fence, sawed a padlock off a door, pried the door open with a pole, and took 10 pounds (4.5 kg) of enriched uranium submarine fuel. The theft was not discovered for twelve hours—and then only because the thief had left the door open. Not knowing how to market the material, which was not enriched enough to serve as bomb fuel, he kept it for six months before being arrested.[19]

Highly radioactive uranium being loaded up for removal to a safe facility

The looseness of borders within the former Soviet Union was demonstrated in a report of 13.2 pounds (6 kg) of HEU that had entered Turkey. It was alleged to have come from Kazakhstan, through Russia, into Chechnya, and then through Georgia before coming into Turkey, where it disappeared.[20]

In June 1997, the U.S. Customs Service arrested two Lithuanians after they offered to sell undercover agents forty Bulgarian shoulder-fired anti-aircraft missiles and tactical nuclear weapons. The two claimed to represent a Colombian drug organization. Neither the missiles nor the nuclear weapons were ever found.[21]

Observers have said that theft of nuclear material is easy in the former Soviet Union because of very lax storage, security, and accounting. Some storage facilities protect radioactive materials with nothing more than a simple padlock. Guards are often young soldiers earning less than $50 per month and therefore very susceptible to bribes. Given the tremendous wealth and reputed brutality of the Russian mafia—supposedly far worse than its Italian and American counterparts—access to uranium and other deadly substances is probably very easy. Given the very high value and easy accessibility of such substances, theft seems virtually inevitable.

Russian officials have admitted that no one in the former Soviet republics really knows how much nuclear material is out there. Accounting was never very accurate, so if some is missing, it is possible that no one will even notice.[22]

The total global underground traffic in radioactive materials, therefore, is all but impossible to estimate. The minor amounts that turn up in arrests and border seizures may only be the tip of a monstrous iceberg. The smugglers that get caught are assumed to be the least clever and most clumsy. In all likelihood, successful smugglers, if any, stayed out of Western Europe and went directly from Russian territory to countries less well protected by the law and from there to the hand of the buyers. Given the worldwide success of drug-smuggling operations—less than 5 percent of worldwide narcotics production is seized—there is every reason to assume criminals have had similar success in the smuggling of radioactive materials.

Unfortunately, the illicit traffic in radioactive materials is not destined for responsible hands. The only possible buyers are terrorists, crime organizations and rogue regimes, and their intentions cannot be good.

Nuclear technology is also leaking out of the post-Soviet republics. An estimated 2,000 scientists are considered capable of designing an atomic weapon. As the economies of their new countries collapse, they are tempted to look for salaries that might pay more than a few hundred dollars per year. Several scientists have been caught trying to leave the country to take jobs in North Korea, and several are known to be working in Iraq.

The Potentials for Terrorism

In 1997, the Committee on Nuclear Policy, organized by the Henry L. Stimson Center, conducted a poll on American fears of nuclear terrorism. The poll found that 76 percent of Americans thought it likely that the United States could be attacked by terrorist groups who smuggle nuclear bombs into the country. Only 8 percent were only slightly or not at all frightened of the possibility.[23]

The potential use of nuclear materials puts unprecedented terror in terrorism. A radiological weapon could be as simple as a few pounds of spent nuclear fuel tossed into the winds over Manhattan, rendering much of New York City immediately, and perhaps permanently, uninhabitable. If the bomb that exploded in the World Trade Center in 1993 had contained a handful of radioactive materials, that building and its surrounding area, including the Wall Street stock market, might have to be sealed off for hundreds of years.

An atomic bomb in the hands of terrorists is by no means an impossibility. The seizures of plutonium and enriched uranium in Europe are evidence that nuclear fuel is available on the black market. Expatriate Russian scientists are selling rogue regimes the latest technology. Between the fuel and the technology, a crude nuclear bomb is a definite possibility. The delivery system that brings it to a target city could be as simple as an airline jet that explodes before its cargo goes through customs. It might also follow the route of the hundreds of tons of illicit narcotics that cross U.S. borders every year.

At least one terrorist group was known to be attempting to develop some kind of nuclear weapon—a Japanese religious sect accused of releasing a lethal nerve gas in a Tokyo subway. The attack killed twelve people and wounded 5,500. Investigations revealed that the group, called Aum Shinrikyo, had made extensive efforts to buy materials needed for chemical, biological, and nuclear weapons. The group had hired at least two Russian nuclear scientists. Among the equipment they may have bought in the United States was a laser system that can be used to precisely measure the plutonium core of a nuclear bomb. Although it would have been illegal to sell the system to Libya, Iran, or other countries with nuclear aspirations, its sale to Japanese companies was perfectly legal. The sect had also tried to buy nuclear materials from sources in Russia. U.S. Senator Sam Nunn of Georgia, a member of the Senate Subcommittee on Investigations, said that the sect's activities were "a prime example of what I believe to be our greatest national security concern in the years ahead."[24]

The mere threat of such an atomic attack by an explosive or radiological weapon could be enough to blackmail a government, set off a stampede of evacuation, and send a local or even a national economy into a tailspin.

When FBI Director Louis Freeh talked with President Lech Walesa of Poland, Walesa said that his biggest nuclear fear was not of a bomb or a terrorist threat but of a criminal organization. He said that if they threatened to dump highly radioactive materials in Warsaw's water supply, he could do nothing to stop them besides give in to their demands.[25]

Some specialists in terrorist tactics think it unlikely that a terrorist group or rogue regime would attempt such an attack or threat. While the bombing of an airliner or

While president of Poland, Lech Walesa (left), here meeting with U.S. President Clinton, raised concerns about radioactive contamination in his country.

train station might provoke mere economic sanctions, the seriousness of an atomic attack would invite a military reprisal that the regime would not survive. The attack would not be worth the consequences.

Likewise, criminologists theorize that the Russian mafia might forego the opportunity to deal in atomic weapons. Its less dangerous areas of business, such as drugs and extortion, are turning a fine profit. Dealing in atomic weapons would draw too much attention and law-enforcement activity.

Problems with Enforcement

Global organized crime has outpaced law enforcement not only in activity but in sophistication and flexibility. While the world's nations still adhere to centuries-old concepts of borders and national laws, organized crime has changed with the times. It crosses borders almost at will. It shifts activities to countries where legal systems are most conducive to illegal enterprises. Its billions of dollars, rubles, francs, yen, pounds, and pesos fly around the world at the speed of light while police agencies have to stop at borders to work through legal channels, diplomatic protocol, political considerations, lethargic bureaucracies, and constitutional rights. A criminal can cross a border far more easily and legally than a law-enforcement officer can. Although organized crime has become a vast international network, there is no international police force to combat it. There are nuclear criminals but no nuclear police.

An incident in Switzerland illustrates how the far the law lags behind nuclear crime. Four men were arrested there just after allegedly selling 11 pounds (5 kg) of plutonium to a Libyan agent, who presumably took his purchase to Libya. The men were released because Switzerland has no law against trafficking in nuclear materials.[26]

When undercover police in Munich trapped a would-be seller of uranium coming out of Russia in 1994, government officials criticized the police. Undercover plans to lure and trap dealers of nuclear materials were actually creating a market for such materials. The Russian mafia, it was said, perceived the undercover police as a lucrative market. The perceived demand then created the supply to satisfy it. Unfortunately, once the supply system was created, it would seek out a real market.

More likely, however, the demand already existed. With plutonium fetching a price of $1 million dollars per 2.2 pounds (1 kg), the supply will surely find the buyers.

Only major changes will make it possible for law enforcement to keep up with the nuclear underground. Never before have criminal organizations had as much money and mobility, and never before have they trafficked in such deadly products.

In 1993 the CIA established the Nonproliferation Center (NPC), an organization dedicated to gathering information on illegal nuclear traffic and staffed with 100 personnel from various federal agencies. The NPC gathers information on worldwide acquisitions of weapons, technologies, components, designs and military systems that are part of weapons of mass destruction. These weapons include chemical and biological weapons. Although the NPC has no powers of enforcement, it serves as the world's central and most comprehensive source of information on proliferation.[27]

With information provided by the NPC and other sources, the United States and the United Nations, through the IAEA, can seek more specific information, formulate policies, begin negotiations, and work toward resolving the problem. Unfortunately, if a country succeeds in developing nuclear weapons, there is little that anyone can do to convince that country to voluntarily forsake them.

Although the CIA is not a police agency with powers of arrest or enforcement, former President George Bush authorized the agency to disrupt the illegal supply of dangerous weapons or their components in foreign countries. This applied to movements by organized crime as well as national governments. Any actions taken by the CIA have been kept secret, although an official in Washington has revealed that the CIA has used its new authority. This powerful authority, however, only makes it more difficult, not impossible, to produce or trade in nuclear weapons and materials.

The FBI has also become involved, though by no means in proportion to the size of the problem. The bureau opened a small office at the U.S. embassy in Moscow to gather information and to train local law-enforcement officials. In 1994, the U.S. government agreed to send $15 million to Russia for the purpose of fighting crime there.

The International Criminal Police Organization, more commonly known as Interpol, is a private, nongovernmental organization that keeps an eye on international nuclear movements. Interpol is not a police organization, however. It has no authority to enforce laws or arrest criminals. As a network involving more than 160 countries, most recently Russia and Ukraine, Interpol only serves as a central source of information for national law-enforcement agencies. Because its information is available to any agency in any country, many countries, in the name of human rights, prohibit their agencies from giving Interpol information about citizens. In some cases, however, Interpol can make information available more easily than it would be through diplomatic channels.

The creation of the European Economic Community has made it very easy for people and products to cross borders. Unfortunately, law-enforcement agencies have not kept up with this new mobility. Few laws cross borders. Extradition of known criminals from one country to another is always difficult and often impossible. In a situation that transcends borders, the police in various countries cannot even agree to communicate on a common radio frequency, let alone work under a common set of laws.

Similarly, the breakup of the Soviet Union created a lot of new borders between countries that have inadequate legal systems, pervasive corruption, and no infrastructure for border control. The most common passport in these new nations is still that of the Soviet Union—a country that no longer exists.

Nonproliferation Policies

Nonproliferation is an integral part of U.S. foreign policy. The Arms Control and Disarmament Agency (ACDA) works with the State Department to develop policies that minimize stockpiles of nuclear weapons material and ensure that such materials are used for peaceful purposes only. The ACDA develops strategies for international cooperation, policies for the control of the production of plutonium and HEU, and works with the IAEA to strengthen its effectiveness.

Diplomacy and national policies have been extensively but not perfectly effective. To prevent, stop, or counteract proliferation through more direct and specific means, the United States has founded the Counter-Proliferation Initiative (CPI). This U.S. Department of Defense program is a strategic set of military plans and capabilities aimed at either interrupting nuclear, chemical, or biological weapons programs or defending against attack by such weapons in war.

With $400 million earmarked for the CPI, the Pentagon has identified several areas in which the military, including NATO, can prepare to face rogue owners of weapons of mass destruction. This new "mission area" of the U.S. military is supposed to be strong enough to handle two regional wars at the same time. Among the capabilities of the CPI are:

- Improved strategies for military operations against countries with weapons of mass destruction
- Special intelligence-gathering to find out the nature, extent, purpose, and ambitions of weapons programs
- Passive defense capabilities such as protective suits, shelters, vaccines, and antidotes to reduce the effects of nuclear, chemical, or biological attack on civilians and military forces
- Active defense capabilities, such as anti-missile systems and anti-contamination measures
- Inspection and monitoring to detect weapons programs so that they can be eliminated before they are used
- Export controls to prevent and detect transactions that contribute to weapons programs
- Tactics for action against weapons programs[1]

In the days of the Cold War, the United States countered Soviet nuclear power with deterrence—the threat of atomic response to atomic attack. The CPI is a non-nuclear strategy against nuclear attack. The big advantage of this strategy is that it is more usable and thus a more credible threat. If would-be proliferators realize that their nuclear weapons programs face likely destruction, they will be less likely to go to the trouble and expense of starting the program. Likewise, non-nuclear countries will have more confidence in a usable, non-nuclear counter-proliferation strategy. They will believe that the United States or NATO can and will protect them from nuclear attack. The CPI, then, is a message to peaceful countries and new nuclear powers alike: If you build a nuclear arsenal, you will probably lose it; if you don't build a nuclear arsenal, you can count on deterrence from the United States.

Representatives of the Arab nations conferring at the United Nations during a 1995 conference on the Nonproliferation Treaty. Cooperation among nations is key to controlling the current nuclear threat.

International Cooperation

To control the proliferation of nuclear weapons, the United States and other countries around the world must engage in unprecedented international cooperation. The borders crossed so easily by criminal and terrorist groups and the agents of rogue nations will have to allow national and international law-enforcement agencies to cross as easily. Though governments may have differences of political and economic opinion, they will have to agree that nuclear proliferation threatens them all.

International corporations will have to cooperate too. No matter where they are based, companies that produce the components and ingredients of nuclear weapons and delivery systems will have to restrict their sales to legitimate buyers.

The number of countries with the capability of developing nuclear technology is growing quickly—more quickly than actual proliferation. The reason is that the decision to "go nuclear" is decided more by politics than by ability. Before making a decision one way or the other, countries tend to consider whether nuclear power will be to their advantage. That decision depends on how easily and economically they can develop a weapon and what the rewards and penalties will be.[2]

To be effective, a nonproliferation policy, national or international, must increase the disadvantages of proliferation and the advantages of nonproliferation. Policy analysts have outlined four basic aspects of such a policy. It would create obstacles, impose disincentives, offer rewards, and reduce perceived needs for nuclear weapons.[3]

Obstacles to Proliferation

Blocking access to weapons, fuel, and components will slow the spread of nuclear weapons. It will also make weapons-development programs take longer, making it easier to stop them through diplomatic efforts.

Among the possible obstacles are secrecy and export controls. Secrecy restricts the flow of technical information and export controls on products that can be used to make weapons. Secrecy must include preventing or discouraging experts from giving assistance to potential proliferators. Export controls must include laws restricting the products that can be exported. To extend controls, governments can discourage exports from other countries through diplomatic efforts or by threatening economic sanctions or even military intervention.

The United States has several laws in place to restrict exports.

- The Atomic Energy Act of 1954 controls the kinds of products and information that can be exported.
- The Nuclear Nonproliferation Act of 1978 requires IAEA safeguards as a condition for export of nuclear fuels and reactors. It also strengthens international controls over the transfer of nuclear technology and U.S. export of dual-use products.
- The Export Administration Act of 1979 requires licenses for controlled products that can be used for nuclear, chemical, and biological weapons, including dual-use products.
- The Arms Export Control Act of 1976 requires licenses for the export of items covered by the International Traffic in Arms Regulations and the U.S. Munitions List.

Besides these unilateral U.S. policies, various international agreements restrict, oversee, or discourage nuclear proliferation. As explained in chapter 4, these include the Nuclear Nonproliferation Treaty (NPT), the Nuclear Suppliers Guidelines (NSG), the Missile Technology Control Regime, and various regional agreements.

The United States also creates obstacles to proliferation through various kinds of pressure. Among the pressure tactics are the denial of "most-favored-nation" trade status (which keeps tariffs low), denial of economic aid, denial of permission to buy other U.S. arms, blocking of international financial transactions, suspension of aircraft-landing rights, and blocking of imports.

In the rare case of an urgent need to block the development of a nuclear weapon by a belligerent nation, the United Nations, the United States, or some other country may opt for military intervention. War, however, is a desperate step, a dangerous and unpredictable solution. It marks the failure of all other tactics. The world is far better off strengthening the power of other options than relying on force.

Disincentives

Disincentives are reasons that a country might choose not to pursue proliferation. Essentially, they are threats.

Economic disincentives threaten sanctions against a would-be proliferator. The United States, being such a large importer of foreign goods and exporter of high-technology goods, has this kind of power. It may restrict imports from the proliferator, or it may prevent certain products from going to that country.

Diplomatic disincentives may threaten to create treaties or military assistance with neighbors or enemies of the proliferator, counterbalancing the advantage that nuclear power might offer.

Military disincentives can threaten to cut off military aid or even carry out sabotage or invasion.

Disincentives work better with some countries than with others. Effectiveness is much greater if the whole world cooperates. U.S. economic sanctions or military restrictions have little effect if other countries are willing to supply what the United States cuts off. Also, the threat of military attack may not bother certain tyrannical governments that care little about how their populations might suffer as long as their regime survives.

When India officially joined the nuclear club in 1998 by testing five nuclear weapons, it did so despite several disincentives. Under the 1994 Nuclear Proliferation Prevention Act, the United States was supposed to cut off all government aid to India and all sales of military equipment. Japan and Germany were also considering cutting off aid. India was expecting to receive $155 million from the U.S. Agency for International Development and $575 million in credits from the U.S. Export-Import Bank, much of which was to be spent on planes, power plants, and other products made by U.S. companies.[4] India could also expect to lose $41 million in licenses to buy U.S. military equipment.[5] The United States was also to oppose lending by the World Bank and the International Monetary Fund (IMF). Although the IMF had no programs with India at the time, the World Bank was expected to lend India $3 billion in the next year. India apparently felt that either the aid would continue or that its loss was worth the boost in military power and national self-esteem.

Rewards

Various policies reward countries for abstaining from nuclear military power. The NPT offers financial rewards, technical assistance, and exemptions from export controls on dual-use products. These rewards make it easier for a country to run a safe, economical, and efficient nonmilitary nuclear program.

Foresaking nuclear military power also rewards a country with the possibility of greater military security. Potential enemies are more willing to sign peace agreements. They are also less likely to develop their own nuclear, chemical, or biological arsenal. Regional security arrangements are more possible and more secure.

The potential of these rewards is effective only if important countries and organizations such as the United States, the United Nations, and the IAEA actively encourage them.

Reducing Perceived Needs for Nuclear Weapons

Countries and organizations can devise more effective antiproliferation policies by studying the motives of potential proliferators. Very often, international policies can be adopted or adjusted to reduce a country's perceived need for nuclear weapons.

If the reason is fear of an enemy's or neighboring country's military power, perhaps the solution is a bilateral or regional treaty. Or perhaps the country would feel more secure if it had a mutual-defense treaty with the United States, NATO, or another major power.

Potential proliferators might also be persuaded that nuclear weapons really are not all that useful. Since the nuclear explosions that ended World War II, the world has maintained a fifty-year tradition of non-use. The next country to use nuclear weapons will surely suffer serious reprisals, probably including military attack by more powerful nations.

Countries with nuclear weapons might also promise to arm the neighbors of potential proliferators, thus eliminating the advantage of acquiring a weapon.

Again, this strategy, as the others mentioned earlier, will be effective only to the extent that it is used in cooperation with many other countries and organizations and in coordination with other strategies.

U.S. foreign policy can get risky when it tries to hinder proliferation among regional enemies. One way to discourage the building of a nuclear arsenal is to offer protection to the country that feels it needs nuclear weapons. The United States might offer protection to that country's enemies instead, making a small nuclear arsenal ineffective. To use it would beg reprisal from the powerful U.S. military.

On a broader scale, the United States could extend its nuclear security blanket over an entire region, threatening reprisal against any country that launches a nuclear weapon. With such a security arrangement in effect, countries in that region would not feel the need to develop their own weapons. To use the words of a columnist with *The Washington Post*, guaranteed U.S. protection would "smother" the regional need for such weapons.[6]

Such promises of protection, however, can lead to disaster. The United States could become involved in a regional conflict that does not threaten U.S. territory. The conflict could even become nuclear. Would Americans support U.S. involvement in a war far from home—nuclear or conventional? Quite possibly not. Unable to carry out its promise of protection, the United States and its counterproliferation policies would lose credibility. Agreements in other parts of the world could crumble, opening the gate to regional nuclear arms races.[7]

The United States has already found itself within range of at least one potential nuclear crossfire. In the Korean situation of the mid-1990s, the United States was obliged to protect South Korea. Thousands of American soldiers were stationed there not only as a defense force but also as a "trip-wire" that would guarantee U.S. involvement in any Korean war. When North Korea appeared to be developing a nuclear arsenal, the United States faced a difficult choice. It could supply South Korea with nuclear protection, thus deterring a nuclear attack by North Korea, or it could pull out and leave South Korea to its own defense. To deter nuclear weapons with nuclear weapons is, of course, to risk nuclear war. To pull out would encourage South Korea to build its own nuclear arsenal.

The possibility of getting involved in someone else's war is not the only calamity that could result from U.S. counterproliferation strategies. Despite U.S. guarantees of protection, some countries could well go ahead and develop nuclear weapons anyway. As demonstrated by the success of programs in Iraq and North Korea, rogue nations can succeed in developing weapons secretly and independently. The result of U.S. policy suppressing proliferation among "good" nations while "bad" nations proceed with nuclear programs could result in a lopsided nuclear balance. Peaceful nations will not have nuclear weapons, while aggressive nations will have nuclear arsenals. The United States could be the only nuclear power capable of maintaining a balance.[8]

The Use of Force

One of the biggest questions facing U.S. and international policy-makers is whether to use force to prevent proliferation. While the NPT prohibits the development of nuclear weapons among its signers, the treaty offers no policy or procedure for enforcement.

Under international law, a single country cannot unilaterally decide to invade another country because it has—or may have or may soon have—a nuclear weapon. Still, it is not hard to imagine a situation in which nonproliferation efforts have failed and an enemy of the United States develops a nuclear capability. Likewise, regional enemies, such as Pakistan and India, may perceive a threat in a neighbor's nuclear armament.

When, then, is it permissible or advisable to launch a military attack to destroy the would-be proliferant's nuclear infrastructure? A paper published by the Naval War College argues that the the United States should have a policy and the capability to use military force as a last resort.⁹

The paper points out the many problems with such a policy. Would it make the United States look like it did not respect international law? What would happen if the United Nations specifically prohibited such an attack? Would the U.S. policy be seen as a precedent for other nations that would like to attack a neighbor with a nuclear weapons program? Would the international nonproliferation regime collapse as the attack proved that the NPT was a failure? Should military counter-proliferation be "legalized" through international diplomacy even if it gives all countries the right to choose that alternative? As with almost all questions about nuclear weapons policies, these have no simple answers.

U.S. Anti-Proliferation Organizations

The United States has proceeded to fight proliferation on two fronts: the diplomatic and the military. Diplomatic efforts try to prevent proliferation or reverse its progress. The experience in North Korea would be a classic example. The NPT and other treaties are also diplomatic efforts at preventing proliferation.

Diplomatic efforts have been largely successful. Most countries that could develop nuclear weapons have chosen not to. Very few are really trying to develop such weapons, and they're having a hard time of it. Australia, Canada, and the non-nuclear countries of Europe, Japan, and South Korea have decided not to use their nuclear expertise to build weapons. Argentina, Belarus, Brazil, Kazakhstan, South Africa, and Ukraine have all abandoned their nuclear programs. Only the handful of rogue nations have dreams of becoming nuclear powers.

As technology improves and nuclear energy becomes more common, plutonium stockpiles grow. As nuclear materials and technology become more commonplace, a black market for nuclear materials and equipment can take hold. Once a market of sellers and buyers develops, the likelihood of proliferation increases. Success in just one rogue nation is enough to cause terrible problems.

The United States, then, has two approaches to nonproliferation. One strives to prevent proliferation. The other plans to counteract it when it happens.

In the End

Only one thing is certain as the world confronts the inevitability of further proliferation: the situation is unlike any in history, and policies unlike any others will have to be devised. Nuclear technology is not likely to go away, nor will the commonly perceived need for stronger security. The world has not seen its last war. The only question is whether the wars of the future will exercise the nuclear option.

To proceed into a new nuclear century without an established, effective anti-proliferation plan and policy is not a good idea. Current policies have worked to a great extent, but many troublesome situations threaten their effectiveness. Global crime organizations are finding it easy to evade the rules of nonproliferation. Technology is making it easier to devise weapons. International transportation is facilitating the movement of people and cargoes. International finances enable electronic funds to reach recipients almost anywhere almost instantly. The nuclear industry is producing ever-greater quantities of very deadly radioactive materials, including potential bomb fuel. In many ways, the world is coming closer to international

This child's scars are a reminder of Hiroshima in 1945. Though further proliferation is probable, the world can hope that past mistakes will not be repeated.

anarchy, with governments unable to control factions and forces within their borders.

The countries and organizations of the world clearly need to work together to devise a unified policy broaching all of these elements of nuclear proliferation. Perfect effectiveness is probably not possible, not until the last bits of plutonium and enriched uranium have withered away. Increased effectiveness, however, is essential. With that and a little luck, the world just might survive long enough to reach a state beyond violence.

Glossary

ACDA Arms Control and Disarmament Agency

atomic bomb a fission weapon that brings nuclear fuel to critical mass to start a chain reaction that results in a nuclear explosion

biological weapon a weapon that releases harmful, often deadly, microorganisms or viruses

breeder reactor a nuclear reactor designed to create more plutonium than it consumes

chain reaction the process by which the splitting of one atom causes the splitting of several other atoms, which in turn split more, releasing a tremendous amount of nuclear energy

chemical weapon a weapon that releases toxic chemicals

conventional weapons weapons that are not nuclear, chemical, biological, or radiological

core a) the fuel and detonation mechanism of an atomic bomb b) the fuel in a nuclear reactor

CPI Counter-Proliferation Initiative

critical mass the minimum quantity of nuclear material needed to start a chain reaction

dual-use products or equipment that can be used to manufacture nuclear weapons or delivery systems but were designed for other industrial uses

fissile material material that is capable of undergoing fission

HEU Highly Enriched Uranium

highly enriched uranium uranium that is about 20 percent pure U-235. It is used as a fuel for nuclear bombs. It can be made useless for bombs if it is mixed with other radioactive materials, including U-238.

hydrogen bomb a fusion weapon far more powerful than an atomic (fission) weapon. The explosion is caused by a smaller atomic explosion that forces the nuclei of pairs of hydrogen atoms to combine, releasing a tremendous amount of energy.

IAEA International Atomic Energy Agency

low-enriched uranium uranium used to fuel nuclear reactors. It usually contains 2 to 5 percent U-235, with most of the remainder being U-238.

MIRV Multiple Independently targeted Reentry Vehicles, a missile warhead system that separates to hit several targets

MOX Mixed Oxide Fuel, a mixture of uranium oxide and plutonium oxide used as a reactor fuel. MOX is produced so that plutonium may be "burned" to create energy.

MTCR Missile Technology Control Regime, a nonproliferation agreement among suppliers of missile components

NPT Nuclear Nonproliferation Treaty

nuclear (or atomic) reactor a device containing nuclear fuel—usually low-enriched uranium—which, through a controlled chain reaction, releases heat and radiation

nuclear weapon (bomb) an atomic (fission) or hydrogen (fusion) bomb

pit the nuclear fuel unit at the center of an atomic bomb

plutonium an element produced in nuclear reactors as U-238 absorbs neutrons to become P-239, P-240, or P-241

radiological weapon a weapon that spreads radioactive materials by explosion, spray, or other means

reactor-grade plutonium plutonium that has spent a long time in a reactor, so that a significant portion of the P-239 has absorbed more than one neutron to become P-240 or P-241. It can be used in a bomb but is less efficient than enriched or weapons-grade plutonium, which is mostly P-239.

reprocessing the breaking up of nuclear fuel rods and separation of plutonium, uranium, and other products that were produced by the chain reaction of the nuclear fuel. Reprocessing separates the plutonium that can be used in bombs.

START Strategic Arms Reductions Treaty

uranium a natural radioactive element that must be purified before it can be used as a nuclear fuel. Its various isotopes include U-235, which can be used to make weapons, and U-238, which can be used as an atomic energy fuel. U-233, 234, and 236 also occur in small amounts.

warhead an atomic bomb that is part of or suitable for a delivery system, such as a missile

weapons-grade plutonium plutonium that has been inside a nuclear reactor for a short time, or that has been separated from other radionuclides, so that it is about 93 percent P-239

weapons-grade uranium uranium that has been separated from other radionuclides so that it consists of over 90 percent U-235

weapons of mass destruction (WMD) chemical, biological, nuclear, and radiological weapons.

Source Notes

1. Executive Order, "Measures to Restrict the Participation of United States Persons in Weapons Proliferation Activities," issued by the White House, September 30, 1993.

Chapter 1

1. Seymour M. Hersh, "On the Nuclear Edge," *The New Yorker* (March 29, 1993), p. 56.
2. "U.S. Security Policy Toward Rogue Regimes," Hearings Before the Subcommittee on International Security, International Organizations and Human Rights of the Committee on Foreign Affairs, House of Representatives (July 8 and September 14, 1993), p. 6.
3. Christopher Drew, "Japanese Sect Tried to Buy U.S. Arms Technology, Senator Says," *The New York Times* (October 31, 1995), p. A5.
4. Bruce W. Nelan, "Formula for Terror," *Time* (August 29, 1994), p. 47.
5. Mark Hibbs, "Those German Headlines," *Bulletin of the Atomic Scientists* (November/December, 1994), p. 25
6. William J. Clinton, "Executive Order: Measures to Restrict the Participation by United States Persons in Weapons Proliferation Activities" (September 30, 1993), p. 1.
7. Hearing Before the Subcommittee on International Security, International Organizations and Human Rights, House of Representatives (April 27, 1993), p. 2.
8. Nelan.
9. George Will, "The Right Bomb in a Suitcase Could

Destroy a City," a syndicated column from the Washington Post Writers Group, Washington, D.C., August 17, 1995.

Chapter 2

1. Associated Press, June 19, 1997, citing *New Scientist* magazine.
2. Associated Press, "Nuclear Inventories," March 13, 1997, citing "Plutonium and Highly Enriched Uranium 1996: World Inventories, Capabilities and Policies" (Oxford University Press and Stockholm International Peace Research Institute, 1997).
3. William J. Broad, "U.S. Sent Ton of Plutonium to 39 Countries," *The New York Times* (February 6, 1996), p. A10.
4. John May, *The Greenpeace Book of the Nuclear Age* (New York: Pantheon Books, 1989), p. 74.
5. Howard Ball, *Justice Downwind* (New York: Oxford University Press, 1986), p. 102.
6. Glenn Cheney, *Journey to Chernobyl: Encounters in a Radioactive Zone* (Chicago: Academy Chicago Publishers, 1995).
7. Kirill Belyaninov, "Nuclear Nonesense, Black-market Bombs, and Fissile Flim-flam," *Bulletin of the Atomic Scientists* (March–April 1994), p. 44.

Chapter 3

1. John May, *The Greenpeace Book of the Nuclear Age* (New York: Pantheon Books, 1989), p. 74.
2. McGeorge Bundy, et al. *Reducing the Nuclear Danger: The Road Away from the Brink* (New York: Council on Foreign Relations Press, 1993) p. 18.
3. "Estimated U.S. and Soviet/Russian Nuclear Stockpiles,

1945–94," *Bulletin of the Atomic Scientists* (November/December 1994), p. 59.
4. Bundy, p. 63.
5. "Estimated U.S. and Soviet/Russian Nuclear Stockpiles, 1945–94," p. 59.
6. Seymour M. Hersh, "On the Nuclear Edge," *The New Yorker* (March 29, 1993), p. 56.
7. "Plutonium Alert," *The Amicus Journal* (Natural Resources Defense Council, Fall 1994), p. 5.

Chapter 4

1. Leonard Spector, "Strategic Warning and New Nuclear States," *Director's Series on Proliferation* (Springfield, VA: National Technical Information Service, August 12, 1994) p. 7.
2. Barbara Crosette, "Treaty Aimed at Halting Spread of Nuclear Weapons Extended," *The New York Times* (May 12, 1995), p. 1.
3. Peter D. Zimmerman, "Proliferation: Bronze Technology Is Enough," *Orbis* (Winter 1994). p. 67.
4. Ibid., p. 76.
5. Chris Hedges, "Iran May Be Able to Build an Atomic Bomb in Five Years, U.S. and Israeli Officials Fear," *The New York Times* (January 5, 1995), p. 10.
6. Alan Cooperman and Kyrili Belianinov, "Moonlighting by Modem in Russia," *U.S. News and World Report* (April 17, 1995), p. 45.
7. For brief information about other arms-related treaties, see the "ACDA Pocket Guide," U.S. Arms Control and Disarmament Agency, Washington, D.C.
8. Richard Jerome, "Bombs Away!" *The New York Times Magazine* (April 3, 1994), p. 46.
9. Charles J. Hanley, "Russia's Atomic Riches: 'Potatoes

Are Guarded Better,' " Associated Press, March 26, 1995.
10. David Kramer, "DOE Still Holds Key Data Confidential," *Inside Energy* (July 4, 1994), p. 1.
11. Mathew L. Wald, "Russia Treasures Plutonium But U.S. Wants to Destroy It," *The New York Times* (August 19, 1994), p. 1.

Chapter 5

1. Claire Sterling, *Thieves' World: The Threat of the New Global Network of Organized Crime* (New York: Simon & Schuster, 1994), p. 48.
2. Jane Perlez, "Treaty to Cut A-Weapons Now in Effect," *The New York Times* (December 6, 1994), p. 10.
3. Michael R. Gordon, "Russia Controls on Bomb Material Are Leaky," *The New York Times* (August 18, 1994), p. 10.
4. Ibid., p. 1.
5. "Russian Army May Be Close to Collapse, a Study Shows," *The New York Times* (February 15, 1995), p. 3.
6. Jane Perlez, "Tracing a Nuclear Risk: Stolen Enriched Uranium," *The New York Times* (February 15, 1995).
7. George Rodrigue, "Visits to Russian Nuclear Sites Alarmed Briton," *The Dallas Morning News* (August 21, 1994), p. 16A.
8. Michael R. Gordon, "U.S. and Britain Relocate a Cache of Nuclear Fuel," *The New York Times* (April 21, 1998), p. 1.
9. Sterling, p. 11.
10. Gordon, "U.S. and Britain Relocate a Cache of Nuclear Fuel," p. 10.
11. Gordon, "Russia Controls on Bomb Material Are Leaky," p. 11.

12. Michael R. Gordon, "Months of Delicate Talks in Kazakhstan Atom Deal," *The New York Times* (November 24, 1994), p. 6.

Chapter 6

1. "Proliferation of Weapons of Mass Destruction: Assessing the Risks" (Washington, D.C.: U.S. Congress, Office of Technological Assessment, August 1993), p. 102.
2. Ibid., p. 103.
3. Ibid., p. 102.
4. George Perkovitz, "A Nuclear Third Way in South Asia," *Foreign Policy* (June 22, 1993), p. 91.
5. Leonard Spector, "Strategic Warning and New Nuclear States," *Director's Series on Proliferation* (Springfield, VA: National Technical Information Service, August 12, 1994), p. 10.
6. Leonard S. Spector, "Repentant Nuclear Proliferants," *Foreign Policy* (Fall 1992), p. 27.
7. Ibid.
8. "Proliferation Threats of the 1990s," Hearings Before the Committee on Governmental Affairs, United States Senate (February 24, 1993), p. 29.
9. Kathy Gannon, "Pakistan Prepared to Conduct Own Nuclear Test," Associated Press, May 13, 1998.
10. Tim Weinder, "Indians Risk Invoking U.S. Law Imposing Big Economic Penalties," *The New York Times* (May 12, 1994), p. 1.
11. "Pakistan, Answering India, Carries Out Nuclear Test; Clinton's Appeal Rejected," *The New York Times* (May 29, 1998), p. 1.
12. Seymour M. Hersh, "On the Nuclear Edge," *The New Yorker* (March 29, 1993), p. 53.

13. "Proliferation Threats of the 1990s," p. 29.
14. "U.S., Waving Ban, Will Sell Arms to Pakistan," *The New York Times* (March 21, 1996), p. A4.
15. Perkovitz, p. 92
16. Ted Galen Carpenter, "Learning to Live with Nuclear Proliferation," *Market Liberalism: A Paradigm for the 21st Century* (Washington, D.C.: Cato Institute), p. 259.
17. Yosef M. Ibrahim, "Egypt Criticizes Israel on Nuclear Arms and Palestinian Talks," *The New York Times* (February 20, 1995), p. A3.
18. Paul L. Leventhal, "The Coming Age of Plutonium," *The New York Times* (May 22, 1993), Op-Ed page.
19. Paul L. Leventhal, "The New Nuclear Threat," *The Wall Street Journal* (June 8, 1994), Op-Ed page.
20. Ibid.
21. Ibid.
22. Leonard Spector, *Director's Series on Proliferation*, p. 10.
23. Leonard S. Spector, *Foreign Policy*, p. 33.
24. Leonard Spector, *Director's Series on Proliferation*, p. 11.
25. Paul Leventhal and Daniel Horner, "Peaceful Plutonium? No Such Thing," *The New York Times* (January 25, 1995), Op-Ed page.
26. Leventhal, "The Coming Age of Plutonium."
27. Martin Sieff, "Theft of Plutonium by Terrorists Is U.S. Policy Critic's Worst Fear," *The Washington Times* (October 23, 1993).

Chapter 7

1. "Proliferation Threats of the 1990s," Hearings Before the Committee on Governmental Affairs, United States Senate (February 24, 1993), p. 15.

2. "Saddam's Nuclear Legion: The IAEA's Scoreboard on Iraq's Nuclear Suppliers," *Proliferation Watch* (March–April 1993), p. 2.
3. Leonard Spector, "Strategic Warning and New Nuclear States," *Director's Series on Proliferation* (Springfield, VA: National Technical Information Service, August 12, 1994), p. 13.
4. Ibid., p. 6. *See also* "Proliferation Threats of the 1990s," p. 19.
5. Leonard S. Spector, "Neo-Nonproliferation," *Survival* (Spring 1995), p. 70.
6. *Proliferation Watch*, U.S. Committee on Governmental Affairs (March–April 1993).
7. *Proliferation Watch*, U.S. Committee on Governmental Affairs (January–February 1994), p. 2.
8. Leonard Spector, "Strategic Warning and New Nuclear States," p. 6.
9. IAEA Inspections and Iraq's Nuclear Capabilities (Vienna, Austria: International Atomic Energy Agency, April, 1994).
10. "Russian Nuclear Scientists Working in Iraq," Associated Press, September 26, 1992.
11. *Proliferation Watch*, p. 4.
12. Barbara Crossette, "A Clean Bill for the Iraquis on A-Arms? Experts Upset," *The New York Times* (April 19, 1998), p. 3.
13. Chris Hedges, "Iran May Be Able to Build and Atomic Bomb in Five Years, U.S. and Israeli Officials Fear," *The New York Times* (January 7, 1995), p. 10.
14. David Albright, "An Iranian Bomb?" *Bulletin of the Atomic Scientists* (July–August 1995), p. 23.
15. Ibid., p. 26

16. Steven Greenhouse, "Russia Says Sale of Atom Plants to Iran Is Still On," *The New York Times* (April 4, 1995), p. 1.
17. Hedges, "Iran May Be Able to Build and Atomic Bomb in Five Years, U.S. and Israeli Officials Fear."
18. Chris Hedges, "A Vast Smuggling Network Gets Advanced Arms to Iran," *The New York Times* (March 15, 1995), p. 1.
19. Greenhouse, p. 1.
20. Associated Press, January 12, 1995.
21. "North Korea's Nuclear Ambitions," editorial, *The New York Times* (August 19, 1998), p. A34.

Chapter 8

1. *Global Organized Crime*: The New Empire of Evil, opening remarks, by Arnaud de Borchgrave (Washington, D.C.: Center for Strategic and International Studies, 1994), p. 140.
2. J. Michael Waller, "Russia's Biggest 'Mafia' Is the KGB," *Wall Street Journal Europe* (June 22, 1994), p. 8.
3. "Germany Says 80,000 Tons of Ammuniition 'Missing,'" United Press, February 16, 1994.
4. Claire Sterling, *Thieves' World: The Threat of the New Global Network of Organized Crime* (New York: Simon & Schuster, 1994), p. 199.
5. Ibid., p. 94.
6. Robert I. Friedman, "The Money Plane," *New York* (January 22, 1996), pp. 24–33.
7. Frank Viviano, "Long Smuggling Trail Ends in Polish Valley," *San Francisco Chronicle* (February 22, 1993), p. A-1.
8. Global Organized Crime, p. xi.

9. Associated Press, May 25, 1994.
10. Mark Hibbs, "Plutonium Powder Puzzles Police," *Bulletin of the Atomic Scientists* (September–October, 1994), p. 8.
11. William J. Broad, "Russians Suspect 3 Sites as Source of Seized A-Fuel," *The New York Times* (August 19, 1994), p. A-11.
12. Mark Hibbs, "Those German Headlines," *Bulletin of the Atomic Scientists* (November–December 1994), p. 25.
13. Reuters, August 30, 1994.
14. Associated Press, August 26, 1995.
15. Jane Parlez, "Tracing a Nuclear Risk: Stolen Enriched Uranium," *The New York Times* (February 15, 1995), p. 3.
16. "The World's Worst Nightmare," 60 Minutes, CBS (October 15, 1995), and "The Russian Connection," *U.S. News and World Report* (October 23, 1995), pp. 56–57.
17. George Rodgrigue, "Visits to Russian Nuclear Sites Alarmed Briton," *The Dallas Morning News* (August 21, 1994), p. 16A.
18. Seymour M. Hersh, "The Wild East," *The Atlantic Monthly* (June 1994), p. 61.
19. Oleg Bukharin and William Potter, "Potatoes Were Guarded Better," *Bulletin of the Atomic Scientists* (May–June 1995), p. 46.
20. Phil Williams and Paul N. Woessner, "The Real Threat of Nuclear Smuggling," *Scientific American* (January 1996), p. 42.
21. Cable News Network, June 30, 1997.
22. Michael R. Gordon, "Russian Controls on Bomb Material Are Leaky," *The New York Times* (August 18, 1994), p. 10.

23. "Public Attitudes on Nuclear Weapons: An Opportunity for Leadership," Henry L. Stimson Center, Washington, D.C., 1998.
24. Christopher Drew, "Japanese Sect Tried to Buy U.S. Arms Technology, Senator Says," *The New York Times* (October 31, 1995), p. A5.
25. Louis J. Freeh, "International Organized Crime and Terrorism: From Drug Trafficking to Nuclear Threats," in *Global Organized Crime*, p. 3.
26. Sterling, p. 217.
27. "Proliferation Threats of the 1990s," Hearings Before the Committee on Governmental Affairs, United States Senate (February 24, 1993,) p. 55.

Chapter 9

1. Stephen A. Cambone and Patrick J. Garrity, "The Future of U.S. Nuclear Policy," *Survival* (Winter 1994–95), p. 73.
2. *Nuclear Proliferation and Safeguards* (Washington, D.C.: U.S. Congress, Office of Technological Assessment, 1977), p. 11.
3. *Proliferation of Weapons of Mass Destruction: Assessing the Risks* (Washington, D.C.: U.S. Congress, Office of Technological Assessment, August 1993), p. 83.
4. "U.S. Is Poised to Slap Sanctions on India Soon," *The Wall Street Journal* (May 13, 1998), p. 12.
5. Tim Weinder, "Indians Risk Invoking U.S. Law Imposing Big Economic Penalties," *The New York Times* (May 12, 1998), p. 1.
6. Stephen S. Rosenfeld, "Bombs for Everyone," *The Washington Post* (June 25, 1993), p. A25.

7. Ted Galen Carpenter, "Staying out of Potential Nuclear Crossfires," *Policy Analysis* (Washington, D.C.: Cato Institute, November 24, 1993), no. 119.
8. Ted Galen Carpenter, "Closing the Nuclear Umbrella," *Foreign Policy* (March–April 1994), p. 11.
9. Frank Gibson Goldman, *The International Legal Ramifications of United States Counter-Proliferation Policy: Problems and Prospects* (Newport, RI: Center for Naval Warfare Studies, Naval Warfare College, April 1997).

For Further Information

Books

Allison, Graham T., et al. *Avoiding Nuclear Anarchy: Containing the Threat of Loose Russian Nuclear Weapons and Fissile Material* (Cambridge, MA: The MIT Press, 1996).

Bailey, Kathleen C., and M. Elaine Price, editors. *Director's Series on Proliferation* (Oak Ridge, TN: Lawrence Livermore National Laboratory, 1994).

Baily, Katherine C. *Strengthening Nuclear Non-proliferation* (Boulder, CO: Westview Press, 1993).

Berger, Melvin. *Our Atomic World* (New York: Franklin Watts, 1989).

Bundy, McGeorge, et al. *Reducing Nuclear Danger: The Road Away from the Brink* (New York: Council on Foreign Relations Press, 1993).

Burrows, William E., and Robert Windrem. *Critical Mass: The Dangerous Race for Superweapons in a Fragmenting World* (New York: Simon & Schuster, 1994).

Cheney, Glenn A. *Journey to Chernobyl: Encounters in a Radioactive Zone* (Chicago: Academy Chicago Publishers, 1995).

———. *They Never Knew: The Victims of Atomic Testing* (New York: Franklin Watts, 1996).

Clausen, Peter A. *Nonproliferation and the National Interest: America's Response to the Spread of Nuclear Weapons* (New York: HarperCollins, 1993).

Costandina, A. *Bombs in the Backyard: Atomic Testing and American Politics* (Las Vegas: University of Nevada Press, 1986).

Forsberg, Randall, et al. *Nonproliferation Primer: Preventing the Spread of Nuclear, Chemical, and Biological Weapons* (Cambridge, MA: The MIT Press, 1995).

Gardner, Gary T. *Nuclear Nonproliferation, a Primer* (Boulder, CO: Lynne, Rienner Publishers, 1994).

Hersh, Seymour M. *"On the Nuclear Edge," The New Yorker* (March 29, 1993).

―――. *"The Wild East," The Atlantic Monthly* (June 1994).

May, John. *The Greenpeace Book of the Nuclear Age: The Hidden History, The Human Cost* (New York: Pantheon Books, 1989).

O'Very, David P., et al. *Controlling the Atom in the 21st Century* (Boulder, CO: Westview Press, 1993).

Pringle, Laurence. *Nuclear War: From Hiroshima to Nuclear Winter* (Hillside, NJ: Enslow, 1985).

———. *Radiation: Waves and Particles, Benefits and Risks* (Hillside, NJ: Enslow, 1983).

Spector, Leonard S. "Repentant Nuclear Proliferants," *Foreign Policy* (Fall 1992), no. 88.

Sterling, Claire. *Thieves' World: The Threat of the New Global Network of Organized* Crime (New York: Simon & Schuster, 1994).

Van Crevold, Martin L. *Nuclear Proliferation and the Future of Conflict* (Toronto: Maxwell Macmillan International, 1993).

Government Publications

"Disposing of Plutonium in Russia," Hearing Before the Committee on Governmental Affairs, U.S. Senate, March 9, 1993.

"Proliferation of Weapons of Mass Destruction: Assessing the Risks," Washington: Office of Technology Assessment, United States Congress, 1994.

"Proliferation Threats of the 1990s," Hearing Before the Committee on Governmental Affairs, United States Senate, February 24, 1993.

Proliferation Watch, all issues. Washington: U.S. Senate Committee on Governmental Affairs.

"U.S. Security Policy Toward Rogue Regimes," Hearings Before the Subcommittee on International Security, International Organizations and Human Rights of the Committee on Foreign Affairs, House of Representatives, July 28 and September 14, 1994.

Internet Sites

The Nuclear Control Institute
http://www.igc.apc.org/nci/index.htm
This site explains the threat of plutonium and uranium and their worldwide trade, as well as the general problem of nuclear proliferation. Includes articles on specific countries and their capabilities. Offers initiatives and publications, and gives suggestions for controlling the current danger. Also provides links to a number of related organizations.

PSR's Nuclear Abolition Campaign
http://www.psr.org/abolitionp.htm
Sponsored by the Physicians for Social Responsibility, this site offers information on nuclear abolition and suggestions for public involvement.

U.S. Nuclear Regulatory Commission
http://www.nrc.gov/
This site offers news and information, a description of the organization, and suggestions for public participation in the regulatory process. Provides references as well as guidelines for preventing radiation exposure and for emergency procedures.

Index

Numbers in *italics* indicate illustrations.

Afghanistan, 32, 72
Africa Nuclear Free Zone Treaty, 49–50
Agreement Between the United States and Belarus Concerning EmergencyResponse and the Prevention of Proliferation of Weapons of Mass Destruction, 50
Agreement Between the United States and Russia Concerning the Safe and Secure Transportation, Storage and Destruction of Weapons and Prevention of Weapons Proliferation, 50
AK-47 machine gun, 6–7
Albright, David, 84
Algeria, 80
Amarillo, Texas, 52, 53
Angola, 32
Anti-proliferation policies, 112–114
Apartheid, 76
Argentina, 70, 80
Arms Control and Disarmament Agency, 105
Arms Export Control Act of 1976, 109
Arms reduction agreements, 52
    costs of, 64
Arniston missile, 76
Atomic bomb, 13–18
    explosions of, 18–19
Atomic capability, 66
Atomic Energy Act of 1954, 109
Atomic terrorism, 22
Atomic weapons, 46
Aum Shinrikyo, 99

Baruch, Bernard, 24, 26, *27*
Baruch Plan, 24, 26
Belarus, 35–36, 50, 58–59, 64
Beryllium, 94–95
Bhutto, Benazir, 72, 73
Black market, 67
Blix, Hans, 83
Brazil, 67, 69
Brezhnev, Leonid, 34
Bulletin of the Atomic Scientists, 95
Bush, President George, 73, 102–103
Bushehr, Iran, 84–85

Cancer, 24
Central Intelligence Agency (CIA), 102–103
Cesium-137, 20
Chain reaction, 15
Chernobyl, 20, *21*
China, 29, 48, 76, 80, 85
    and bomb explosion, 1964, 30
    and Pakistan, 81
    and rogue regimes, 80
Christopher, Warren, 84–85
Churchill, Winston, 27
Clinton, President Bill, *11*, 51, *39*, *100*
    and North Korea, 88
    and START, 42
Cold War, 26–36
Colombian drug organizations, 96
Committee on Nuclear Policy, 98

Comprehensive Test Ban Treaty, 42, 51–52
Contamination, 20, 22
Control system, 17–18
Cooperative Threat Reduction Program, 63
Council for Foreign and Defense Policy, 61
Counter-Proliferation Initiative, 105
  capabilities list, 106
Critical mass, 15
Cuba, 32

Defense, U.S. Department of, 105
de Gaulle, Charles, 67
Delivery systems, 17–18, 45–46
de Mello, Collor, *68*, 69
Disarmament, 40
Disincentives, 110–111
Dismantling process, 52–53
DNA damage, 19
Dole, Robert, 88
Douinreay, Scotland, 63
Dresden, Germany, 19
Drinking water, 22
Dual-use technology, 46

Eastern Europe, 58
Egypt, 41
Eisenhower, President Dwight D., 29
Electromagnetic waves, 13
Energy, Department of, 53
Enforcement, 101–104, 114–115
Environmental contaminants, 20
"Equalizer" factor, 45
Estonia, 94
Eurasia, 91
Export Administration Act of 1979, 109
Extradition law, 103

Federal Bureau of Investigation (FBI), Moscow, 9, 65, 103
Fireball/firestorm, 18–19
Firing device, 16
Fissile material, 15, 17
Force, use of, 114–115
Foreign aid, 65
France, 30, 51
Freeh, Louis J., 9, *92*, 93, 99
Fusion bomb, 17

Germany, 86, 93,110–111
Glasnost, 35
Goiânia, Brazil, 20, 23
Gorbachev, Mikhail, *34*, 35, 57
Greenpeace, 95
Gulf War, 81–82
Gun-assembly technique, 16

Heat release, 18
HEU. *See* Highly enriched uranium
Highly enriched uranium (HEU), 13–14, 16, 50, 59–60
  control of, 22
  in Turkey, 96
Hiroshima, *4*, 16, 19, 24, *25*, *26*, 74, *117*
Hosokawa, Morohiro, 74
House Subcommittee on International Security, International Organizations and Civil Rights, 9
Hungary, 93
Hussein, Saddam, 81, *82*
Hydrogen bomb, 17, 29

Illegal nuclear traffic, 90–104
Immunological illness, 19
Implosion technique, 16
India, 37, 45, 67, 70–71
  and disincentives, 110
  and NPT, 40
Inspection problems, 47

Institute for Science International Security, 40, 84
Institute of Physics and Power Engineering (Russia), 61
International Atomic Energy Agency (IAEA), 14, 43, 78
  and disincentives, 109
  and Iraq, 82–84
International cooperation, 107–108
International law, 114
International Monetary Fund (IMF), 71, 111
International Traffic in Arms Regulation, 109
Iodine-131, 20
Interpol, 103
Iran, 29, 37, 48, 84–86
  and China, 48
  and Israel, 85
  and NPT, 41
Iraq, 8, 36, 47, 79, 81–84
  and China, 48
  and Israel, 81
  and NPT, 41
  Russian scientists in, 98
Iraq Osirak reactor, 85
"Iron Curtain," 27
"Islamic bomb," 37, 84
Isotope, 13
Israel, 67, 73–74, 76
  and Iran, 85
  and Iraq, 81
  and NPT, 40, 41

Japan, 8–9, 74–75, 110–111
  and NPT, 41
  and Tokyo subway bomb, 99

Kahula nuclear facility, 73
Kashir, 71
Kazakhstan, 50, 58–59, 64, 65
  nuclear power of, 35–36
KGB, 90
Khrushchev, Nikita, *31*, 32

Korea, 113
Korean War, 29
Kuchma, Leonid, *39*
Kuwait, 82

Latin America, 49
"Launch on warning," 33
Lebed, Alexander, 62
LEU. *See* Low Enriched Uranium
Leukemia, 19, 24
Levanthal, Paul, 84
Libya, 37, 79
Limited Test Ban Treaty, *51*
Lithuania, 94
"Little Boy" bomb, *16*
Llantos, Tom, 9
Low enriched uranium (LEU), 13, 50
Lugar, Richard, 9, 63

Major, John, *39*
Megatonnage, 9–30
Military intervention, 109–110
Military security, 111
Mines, 18
Missile technology, 48
Missile Technology Control Regime, 48, 109
Mixed oxide fuel, 56
Money laundering, 91–92
"Most-favored-nation" trade status, 109
Multiple Independently targeted Reentry Vehicles (MIRVs), 32
Mutually assured destruction, 32

Nagasaki, 16, 19, 24, 74
NATO, 29, 105–106
Naval War College, 114
Neutrons, 15
1994 Nuclear Proliferation Prevention Act, 71, 110
Nodong I missiles, 86
Nonproliferation Center (CIA), 102

Nonproliferation policies, 105–118
  treaties, 48–52
North Atlantic Treaty
  Organization. *See* NATO
North Korea, 8, 36, 41, 75, 79,
  81, 86–88
  and beryllium, 95
  and inspections, 47
  Russian scientists in, 98
  and Scud-C missiles, 86
Nowy Targ Valley (Poland), 91
NPT. *See* Nuclear
  Nonproliferation Treaty
NPT Extension Conference of
  1995, 74
Nuclear Control Institute, 84
Nuclear explosions, 18–19
Nuclear Fuel (magazine), 62
Nuclear fuel disposal, 54–55, 56
Nuclear materials, 62, 92–98
Nuclear Nonproliferation Act of
  1978, 109
Nuclear Nonproliferation Treaty,
  1970, 10, 39–48, 109, 111
  factors against, 45–48
  and information sharing, 47
  signing of, 36, *39*, 40–41
  success of, 44–45
Nuclear police force, 47–48,
  101–104
Nuclear powers, 1970, 39
Nuclear proliferation, 46–47
  obstacles to, 108–110
  status of, 66–78
Nuclear reactors, 19–20
  spent fuel of, 17
Nuclear scientists, 47
Nuclear submarine fuel, 6
Nuclear Suppliers Group, 49
Nuclear Suppliers Guidelines,
  109
Nuclear technology, 10, 13–23,
  97–98
  definition of, 67
"Nuclear umbrellas," 44

Nuclear waste, 15
"Nuclear-weapon-free zones," 43
Nuclear weapons, 10, 38, 111,
  112–114
  disposal of, 52–56
  non-use of, 44–45
  in post-Soviet countries, 57–65
Nunn-Lugar program, 63
Nunn, Sam, 63, 99

Oak Ridge, Tennessee, 65
Organized crime, 91–92
Osirak reactor (Iraq), 81, 85
Outer Space Treaty, 49
"Overkill," 31
Oxide fuels, 56

Pakistan, 8, 37, 48, 70–73
  and India, 45
  and Iran, 85
  Kahula nuclear facility, 73
  and NPT, 40–41
Pantex complex, 52, *53*
Pelindaba, Treaty of, 50
Pentagon, U.S., 105–106
Perestroika, 35
Plutonium, 7, 13–14, 22, 77–78
  disposal of, 56
  in Japan, 75
Plutonium-239, 13, 20
  half-life of, 54
Poseidon submarine, 32–33
Post-Soviet nuclear powers,
  57–65
Prague, 94
Pressler, Larry, 73
"Principles and Objectives
  for Nuclear
  Nonproliferation and
  Disarmament," 42–44
Protective shielding, 20

Quaddafi, Muammar, 79, *80*

Radiation poisoning, 19

141

Radioactive contamination, 19–22
Radioactive isotopes, 7, 8, 13, 20
Radioactive materials, 7, 8, 20, 22–23
   in Russia, 96–97
   uncontrolled possession of, 92–94
   underground traffic in, 90–104
Radioactivity, 13
Radionuclides, 19
Rarotonga, Treaty of, 49
Reagan, President Ronald, 33–34
Rodionov, Igor, 61
Rogue regimes, 69, 79–89 list of, 81–88
Rokkasho, 75
Russia, 35–36, 58–60, 64
   and arms reduction, 52
   and Iran, 85
   scientists expatriated from, 97–98
   selling of plutonium, 95
   and START, 41–42
   unsafe storage in, 53–54
Russian mafia, 9, 63, 90–104
   and radioactive materials, 92–98
Russian Police College, 9

Scud-C missile, 86
Seabed Arms Control Treaty, 49
Seawolf submarine, 33
Shock waves, 18
60 Minutes, 62
Smuggled goods, 91
South Africa, 67, 74, 75–76
   and NPT, 40
South Korea, 75, 86
South Pacific nations, 49
Soviet Bloc, 27
Soviet-Chinese relations, 30
Soviet "satellite" countries, 58

Soviet Union, 9, 29, 35, 57–58
   breaking up of, 10, 50, 58
   and Cuba, 32
   economic chaos in, 35, 57
   and nuclear weapons, 58–59
   and Russian mafia, 90–104
   uncontrolled borders of, 96, 104
Spent nuclear reactor fuel, 13–15, 17
Stalin, Joseph, 26–27, *28*
Stark, Pete, 78
START. *See* Strategic Arms Reduction Treaty
Stimson, Henry L., Center, 98
Stinger surface-to-air missile, 6
Strategic Arms Reduction Treaty (START), 41–42
Strontium-90, 20
Subatomic particles, 13
Superpowers, 32

"Tactical" warheads, 18
Taiwan, 76
Tbilisi (Georgia), 63
Terrorism, 36–37, 98–104
Terrorists and nuclear arms, 38
Thermonuclear bomb, 17
38th Parallel, 29
Threshold Test Ban Treaty, 51
Tlatelolco, Treaty of, 49
Tokyo subway incident, 8–9, 18–19, 99
Treaties, 38–56
   NPT, 38–48
Trident submarine, 33
"Trip-wire" strategy, 113
Turkey, 96

Ukraine, 50, 59–60, 64
   Chernobyl explosion, 59
   nuclear power in, 35–36
Underground nuclear tests, 37, 51
Unenriched nuclear material, 17

Unexpected acceleration, 81
United Nations, 24, 26
United Nations International Atomic Energy Agency (IAEA), 40–41
United States, 52, 106, 109
    counterprolieration strategies of, 113–114
    foreign policy of, 112–113
    and post-Soviet republics, 63–65
    and START, 41–42
United States Agency for International Development, 111
United States anti-proliferation organizations, 115–116
Uranium, 7, 13–14, 17
Uranium-235, 13, 56
    half-life of, 54
U.S. Export-Import Bank, 111
U.S. Munitions List, 109
U.S.-Russian Agreement Concerning the Disposition of Highly Enriched Uranium Resulting from the Dismantlement of Nuclear Weapons in Russia, 50

Vaccines/antidotes, 106

Walesa, Lech, 99, *100*
Warheads, 30
Washington Post, 113
Weapons disposal, 41
Weapons-grade fuel, 13–14
Weapons inspection, 84
World Bank, 71, 111
Worst case scenarios, 6–12

Yeltsin, Boris, 36, *39*, 57
    and Russian mafia, 91
    and START, 42

143

About the Author

Glenn Alan Cheney is the author of books on Chernobyl, nuclear weapons testing, and several other books on controversial issues. An active member of The Green Party, he works toward the elimination of nuclear power and stronger efforts at arms control. He lives in Hanover, Connecticut, with his wife and son.

10258

327.1
Che   Cheney, Glenn Alan

        Nuclear prolifer-
        ation

DUE DATE   **BRODART**   **07/00   24.00**